The Importance of Soft Skills in Engineering and Engineering Education

Edoardo Rovida · Giulio Zafferri

The Importance of Soft Skills in Engineering and Engineering Education

Springer

Edoardo Rovida
Politecnico di Milano
Milan, Italy

Giulio Zafferri
Officina della Conoscenza
Milan, Italy

ISBN 978-3-030-77248-2   ISBN 978-3-030-77249-9   (eBook)
https://doi.org/10.1007/978-3-030-77249-9

This Springer imprint is published by the registered company Springer Nature
Switzerland AG
The registered company address is: Gewerbestrasse 11, 6330 Cham, Switzerland

Foreword

It is always difficult to write the preface of a book, but it is even more so when the author is a friend who has worked so hard for the university to which I belong and for his Foundation that I preside.

Moreover, I am of a synthetic nature and therefore reluctant to write articulated essays, but the subject dealt with in the book is extremely topical and really very intriguing, and the reading of the drafts passionates me.

For decades, there has been a contrast between "hard skills" and "soft skills", with positions very often more similar to expressions of sport cheer rather than acute reflection on the potential and the need to possess both skills in a balanced way.

Only recently in the technical scientific literature has begun to recognize the importance of soft skills, thanks to the acquired awareness that "being a good technician is not enough to successfully complete a project", on the other hand, even in the literature of social science extraction has begun to consider the relevance of hard skills in a completely different perspective, thanks to the acquired awareness that issues such as digital transformation (only as an example) they must be part of the cognitive baggage of any manager and that it is not possible to successfully manage a project or an enterprise without having a good knowledge of the technical principles that are at the base of the products that are managed.

The advent of "Industry 4.0" in both manufacturing and service industries has finally made it clear to everyone that the border between hard and soft skills is only an artificial border and that any "professional" regardless of the role and the company must master both.

This opens up a not simple problem that is "how to teach soft skills to individuals who by their aptitude have a predilection for hard skills?" and more generally "how to make the teaching of these soft skills attractive in order to maximize the effectiveness of the learning process".

This is the content of the book that deals, in the sequence of chapters, with the individual soft skills, from communication to creativity, providing for each a clear interpretative key and above all an interesting methodological support useful for the trainer.

As President of Fondazione Politecnico, I am therefore grateful to Edoardo and Giulio for this work that provides an extremely exciting analysis perspective on a topic that is increasingly relevant in the training processes and that, as explained in the introduction, became even more in the era of COVID-19.

Andrea Sianesi
Fondazione Politecnico di Milano
Milan, Italy

Acknowledgments

First the authors would like to send a particular warm thanks to Prof. Andrea Sianesi, President of Politecnico di Milano Foundation, for having accepted writing the preface.

In addition, they are also very grateful to both Dr. Alberto Caleca and Dr. Giuseppe Susani for allowing the reproduction, of a large part of Chap. 5, from a publication of the Order of Engineers of Milan.

Finally, they also would like to extend their warmest thanks even to Dr. Giulia Ippolito, for developing the Personal Branding Algorithm, to Dr. Alberta Gianotti, training and coaching senior consultant, soft skills specialist helping people in changing/improving behavioural aspects, as well as Maestro Luciano Beretta and Angelo Beretta, both Art Directors, developing figures and graphics.

Last but not least, author Edoardo Rovida would like to thank very much his wife, Rita Brunetti, for her support. Many thanks, also, to the colleague and friend, as well as co-author, Dr. Giulio Zafferri.

Moreover, even author Giulio Zafferri would like to thank his wife Carla Maria Brivio, his daughter Patrizia and his son Stefano Andrea, and the colleague and friend as well as co-author, Prof. Edoardo Rovida for the support.

Book Purpose

Outline of the soft skill fundamentals aiming at improving new behaviour is called "difference maker".

While the authors are focused on writing this book, the coronavirus tragedy unfolds, spreading death and the premise of an unprecedented economic crisis throughout the world.

We are all experiencing a very painful situation, and the authors wondered if and how reflections on soft skills could offer some interesting insights. Hence, they decided to examine the soft skills discussed in the various chapters, exploring possible links with the current dramatic circumstances.

A few considerations on this "field experience" are given below.

1. *Communication*: in a situation such as this one, correct communications are of the utmost importance, scientifically valid (the "what" of communication) and conveyed in an accessible way to have an instant impact (the "how" of communication). However, the actual situation is often very different. Indeed, we are overwhelmed by a great deal of news (that is even excessive), including fake news that swarms the Internet.

   When we switch on the TV, at any time of the day and on any channel, there are nothing but talk shows with opinion makers, influencers and "spin doctors" saying everything and its opposite (we have even heard people praising COVID because... it would reduce pollution levels!). Fortunately, experts, doctors and scientists also voice their opinions, which should be given more importance. Actually, they should be more "ruthless" and harshly criticize lies.

2. *Courtesy*: never as in these days has respect, both for medical staff and for everyone else, been so necessary. However, TV debates where the exchange of ideas among participants is anything but courteous are rather common.

   Interruptions, hindrances, taunts, if not outright insults, are part of almost every debate.

3. *Flexibility*: humanity has never experienced such a situation before, at least not for a very long time. This is why, as mentioned above, we are swamped by a barrage of communication, even from experts on the subject, information that is often divergent and, at times, even contradictory. In fact, when faced with contradictory opinions, we often shift from a moment of hope to a dizzying loss of it.

   Flexibility and quick assessments make decisions extremely important.

4. *Ethics and integrity*: now more than ever a strong professional ethics is crucial, and healthcare workers are a prime example of this. Their empathy and relational behaviour are a great example of professionalism and ethical conduct for all of us. We have all been moved by the image, shared on the Internet, of the nurse who, clearly exhausted by the gruelling shifts at work, fell asleep on her computer keyboard.

   Unfortunately, the same ethical conduct shown by healthcare workers is not always exhibited by others who often attempt to spread fear, when not outrightly defrauding people.

5. *Interpersonal relationship*: especially in times like these, it is important to discover (or rediscover) interpersonal relationships and affection for those close to us. We can even observe, in a number of cases, the new blossoming of a sense of solidarity that seemed to be forgotten.

   It is, for example, heartwarming to see the intense involvement in voluntary services, both on the part of associations and of individuals who help the most vulnerable people in many ways.

6. *Positivity*: it is important to see things in a positive light and to think that the emergency will come to an end sooner or

later. And especially soft skills can be a great help in thinking positively.

This does not mean being obtusely optimistic ("how bad can it be, it will take care of itself"), but facing the problem with objectivity and, despite taking all the necessary precautions, trying to live "normally" as far as possible.

7. *Professionalism*: healthcare workers provide a clear example that requires no comment and which was underscored in point 4.

   But great professionalism is also required (and often, shown) even by other groups of workers: from shopkeepers to post office clerks, from bank clerks to street cleaners. Their professionalism allows all of us to access essential services.

8. *Responsibility*: we are called to comply with the rules, to consider how our rash behaviour can damage other people. Moreover, the behaviour of those speculating on this situation, those trying to get rich on the distress of others, is beyond comment. In moments like these, our responsibility as citizens and our willingness to help those in need or those who do not have the means and the ability to react, with whatever means we can (and must), come to the forefront. But the sense of responsibility must also involve the people around us: from our neighbours, to our family members. In fact, this period of time spent "in lockdown" can lead to conflict and issues that everyone must try to overcome through interpersonal relationships and a positive attitude.

9. *Teamwork*: here again, the example set by healthcare workers and by those working to elaborate strategies is beyond praise. It is interesting to observe the development of medical equipment, the result of great teamwork of doctors and engineers. In this regard, it is also important to underline the teamwork of healthcare workers from different countries: the help that Italy receives, in terms of both medical equipment and personnel (and, therefore, of professional expertise), from China, Russia and Albania, for example, requires no further comment.

10. *Empathy*: the ability to understand the feeling of others is very important at a time full of difficulties like this. It

is important to be able to help others in their moments of difficulty.

11. *Engagement*: the great difficulties of the moment require a great coordination between the various activities. The ability to engage and the willingness to be involved therefore play a very important role.

12. *Leadership*: at all times of emergency, it is necessary for everyone to carry out their tasks in the best possible way. The figure of the leader is, therefore, of great importance.

13. *Learning to learn*: in a period of emergency like this, the competence of each one is fundamental, in his own work: training and updating in every field are, therefore, activities of great importance.

14. *Problem solving*: the problems at this time are innumerable and continuous: so the ability to identify and solve problems is a fundamental skill.

15. *Digital thinking*: the emergency of today requires to apply the smart working and the online teaching. Therefore, the ability to make optimal use of digital devices is evident.

16. *Planning*: the great problems related to the COVID emergency require to planning means capable of coordinating personal work and work of others, to achieve the objectives.

17. *Result orientation*: in any activity, it is very important to have a clear mental picture of the goal to be achieved. In an emergency time like this, this is particularly important.

18. *Awareness*: it is very important during these times to preserve a balance between opposing moods, such as, for example, between the darkest depression ("we will never get out of this") and inappropriate euphoria ("there's no need to worry, it will be over soon"), or between minimizing risks and psychosis, which can both lead us to make the wrong decisions.

19. *Personal Branding*: it is essential to know oneself and to be aware of the know-how possessed. Particularly at this time, it is essential to have a clear idea of each person's skills to deal with the COVID emergency in the best possible way.

20. *Imagination and creativity*: it is important to let our imagination work to "invent" new activities that can be carried out

while we are forced to stay at home, in order to turn a situation that might feel oppressive into something that, as far as possible, is not totally negative.

Spending long hours and days at home can help us rediscover a new way to relate to others, and also new household activities that have perhaps been forgotten for years.

21. *Human aspects of technology*: there is no need to emphasize the essential role of medical equipment, the result of biomedical engineering that saves numberless lives. Insightfully directed technology has never demonstrated its service to man, and not against man, as in these days.

22. *Talent development*: everything mentioned above would not be possible without developing talent, the main "tool" for any type of progress.

Talent, when well directed and applied, acts as a sort of "glue" between the soft skills we mentioned. As a conclusion to this premise, we can observe that a keen reflection on soft skills can yield some effective ideas to face this emergency.

And, after dealing with some of the main soft skills, the authors, critically examining their experience, formulate some proposals. They could be considered and provide a basis for discussion for their introduction into education programmes in the field of engineering, but not only in it.

And also this is, precisely, one of the aims (if not the basic aim) of this book.

It is interesting to note that an European project called Lever Up has been launched to develop and validate competencies, with particular regard to soft skills [1]. The Politecnico di Milano Foundation and, particularly, one of the authors of this present book is involved in the programme.

The competencies considered in the above project are presented in Table 1.

Table 2 presents the above Soft Skills presented in the book, classified in relation to the "Lever Up categories".

Starting from the above considerations, the general purpose of this book can be described as follows:

Table 1 Competencies included in the "Lever Up Project"

Personal	Methodological	Social	Organizational
(a) Empathy (b) Engagement (c) Flexibility and adaptability (d) Initiative (e) Responsibility	(a) Innovation and creativity (b) Learning to learn (c) Problem solving	(a) Communication (b) Intercultural and diversity (c) Leadership (d) Teamwork	(a) Digital thinking (b) Planning (c) Result orientation

Table 2 Soft skills presented in the book, classified in relation to the "Lever Up categories"

Personal	Methodological	Social	Organizational
(a) Flexibility (b) Responsibility (c) Professionalism (d) Learning to learn (e) Problem solving (f) Digital thinking (g) Planning (h) Result orientation (i) Personal branding	(a) Imagination and creativity (b) Talent development	(a) Communication/ information (b) Teamwork (c) Human aspects of technology (d) Positivity (e) Work ethic and integrity (f) Empaty (g) Engagement (h) Leadership	(a) Interpersonal relationship (b) Courtesy (c) Awareness

To facilitate the possibility of developing personal "ACUMEN", through the improvement of the "INTELLECTUAL FUSION", in order to achieve a special condition of uniqueness and, consequently, become a "DIFFERENT MAKER" kind of person (see Chap. 10).

In order to achieve such a result, it will be necessary to be humble, courageous and determined to develop your personality so as to become the "BEST VERSION OF YOURSELF" at least.

The overall structure of this book is based on one chapter dedicated to each soft skill.

The configuration of each chapter is, in general, as follows:

(a) general principles about the soft skill, e.g. definition, characteristics, fundamental aspects;

(b) some considerations about the important role of the soft skill in daily life; in this sector too, soft skills are highly significant for interpersonal relationships;

(c) the role of the soft skill in the engineering profession: it must be said that, along with hard skills, engineers need a firm foundation in the field of the soft skills too; the engineer is a person who works in contact with many people, therefore, soft skills (e.g. communication, flexibility, interpersonal relationship, positivity and the other topics discussed in the book) play a significant role;

(d) the role of the soft skill in teaching activities and, particularly, in engineering education; indeed, the above-mentioned role of the soft skill in the profession requires consolidated training. It is important to observe that, in addition to teaching soft skills, a teacher must apply soft skills during teaching activities. Such a consideration is true not only for engineering education but also in all fields and levels of education. The aim of this paragraph is to underscore such importance.

Reference

1. www.leverproject.eu. Last visit 25 April 2020

Contents

Introduction

1

Definition of Soft Skills and their important role during one's lifetime.

1.1   General Definitions

Each profession requires the specific knowledge and competencies necessary to perform the relative tasks. Such knowledge and competencies, called skills, can be considered as being divided in two broad categories: Hard Skills and Soft Skills.

The former are the technical competencies, that can be attained through studies and tested by exams; as examples of hard skills, it is possible to consider the ability to formalize a function performed by a technical system; the ability to manage a given software; the ability to calculate the mechanical stress in a given part of a machine; the knowledge of foreign languages; and the ability to utilize machines and devices.

Soft Skills, instead, are personal competencies, related to human personality. They are more difficult to be taught and tested. But they are very important, not only in the Engineering field, but also in all professional fields as well as in everyday life.

© Springer Nature Switzerland AG 2022
E. Rovida and G. Zafferri, *The Importance of Soft Skills in Engineering and Engineering Education*,
https://doi.org/10.1007/978-3-030-77249-9_1

In fact, Soft Skills are related to human behaviour, in all fields of activity: behaviour in relation to human beings and to the specific environmental situation.

Examples of Soft Skills include: the ability to communicate, imagination, leadership, ethics, and the ability working in a team.

1.2   Origins

Soft Skills concept is a very recent discovery, but some ideas of "soft" competencies appeared many centuries ago. The first steps in the development of Soft Skills are, therefore, to be found in the history. For example, the ancient Greek philosopher Aristotle says that efficient communication should be based on three pillars: *ethos* (ethics), *pathos* (emotion) and *logos* (logical rigor). Another very important Greek philosopher, Socrates, claims that the basis of communication is the *maieutikè* (the truth search criteria): the teacher helps the student to bring out the truth. Many centuries later, the very well-known sentence by Erasmus of Rotterdam (1466–1536) [1] "*the best ideas do not come from reason, but from a lucid, visionary madness*" is a primordial idea of Soft Skills. In this sentence the idea that reason is very important is clear, but there is something in addition to reason which can make the behavioural difference.

Furthermore, the well-known French philosopher, Blaise Pascal (1623–1662) [2] recognized two fundamental forms of knowledge, starting from different assumptions, one proposing an *esprit de geometrie* and the other an *esprit de finesse* (spirit of geometry and spirit of intuition). Pascal distinguished, in an initial form, the implicit concepts of hard and soft competencies, and highlighted the important role of the "soft". Pascal says that the spirit of geometry is not enough to completely understand reality, because it is not able to get to the foundations of the human being. Each science, without consideration for the human being is, in Pascal's opinion, not useful in general. However, the spirit of intuition, without spirit of geometry is weak and fails to descend to the deepest and truest principles of the human being.

Consequently, the spirit of intuition and the spirit of geometry are both equally necessary for complete knowledge and, ultimately, for a complete human being.

Therefore, based on the integration of the two "spirits", Pascal's thought contains an embryo of the concept of "Intellectual Fusion", that will be a fundamental aspect of soft skills.

1.3   Characteristics of Soft Skills

Nowadays a knowledge of soft skills is particularly important during recruitment processes. As observed above, while it is easy to upgrade hard skills, it is not, in a similar way, easy to upgrade soft skills. Consequently, problems in teaching soft skills can produce problems in the integration of the competencies and, consequently, in team working.

In this book, with reference to the bibliography [3] and to their teaching and research experience [4], the authors will deal with the soft skills listed in Table 1.1.

Soft Skills based on one of the authors' experience with project Lever Up of the Politecnico di Milano Foundation have been added to the Soft Skills listed in Table 1.2.

With reference to point m) of the Table 1.1 of particular interest and use is the anticipation of some considerations about the link between hard/soft skills and the parts of the human brain.

Hard skills (HS) are related to the left part of the brain and are linked to the Intellectual Quotient (IQ), while the soft skills (SS) are related to the right part of the brain and linked to the Emotional Quotient (EQ). The integration of IQ and EQ has IF (Intellectual Fusion), as a result:

$$IQ \times EQ = IF, \text{ requirement for each professional}$$
$$\text{to achieve the "role" of "difference maker".}$$

Another interesting analytical relation is as follows:

$$IF \times XQ = QS, \text{ quotient of success}$$

Table 1.1 The soft skills considered in this book

Soft skills	Explications
(a) Communication capability, based on verbal, non-verbal and ability to listen	The ability to structure each transmission of information (particularly, but not only, technical) by applying the following steps: (a.1) Identification of the receiver's initial knowledge of the communication (starting point) by considering the hard and the soft components (a.2) Identification of the upgrading of knowledge to be attained by the receivers of the communication (arrival point) by considering the hard and the soft components (a.3) Identification of the knowledge to be transmitted (what to transmit) and consequent feedback to be received (a.4) Determination of the communication means necessary for the transmission (how to transmit) (e.g. text, voice, picture, by regarding hard and soft components)
(b) Courtesy/kindness	(b.1) The ability to win the respect of all people you come into contact with
(c) Flexibility	(c.1) The ability to shift thought and action course according to the environmental changing
(d) Ethics and integrity	(d.1) The quality of being honest and of having strong moral principle (d.2) The state to consider that technical products could cause damage to their users in some cases (honesty in engineering deontology)
(e) Interpersonal relationship	(e.1) The ability to create a pleasant social commitment (e.2) The ability to realize good relationships with everybody

(continued)

Table 1.1   (continued)

Soft skills	Explications
(f) Positivity	(f.1) The ability to be optimistic and to transmit optimism to colleagues
(g) Professionalism	(g.1) The ability to behave in a way that is congruent to the requirements of the specific work/social environment (g.2) The ability to update professional knowledge, in relation to the change of the market's/environment's requirements
(h) Responsibility	(h.1) Self-discipline and conscientiousness
(i) Teamwork	(i.1) The ability to cooperate and to contribute so that all members of the team perform the required task
(j) Imagination and creativity	(j.1) As said by A. Einstein "imagination is more important than knowledge. Knowledge is limited. Imagination encircles the world"
(k) Awareness	(k.1) How to develop self-knowledge and self-evaluation skills
(l) Human aspects of technology	(l.1) A clear idea that technology is for the mankind and not mankind for technology (l.2) The ability to evaluate positive and negative effects of technical products on the user
(m) Talent development	(m.1) The ability to reach a complete personality by a balance between the intellectual quotient and the emotional quotient: the result will be intellectual fusion

Table 1.2 Soft skills derived from the experience in the lever up project

Soft skills	Explication
(aa) Empathy	Understand the feelings of others, step into their shoes, interpret their behaviours
(bb) Engagement	Involve others in sharing new ideas and new initiatives, emotional commitment in being engaged
(cc) Learning to learn	Powerful mental tool stimulating motivation and desire to acquire new knowledge and know-how in order to update and implement personal and professional experience
(dd) Problem solving	It is all about using logic and imagination to recognize problems, analyze them and define targeted strategic solutions
(ee) Digital thinking	Develop awareness of the opportunities offered to people with digital mind-set to understand the world and believe in the power of data
(ff) Leadership	Inspire and have a positive influence on the people behaviour by channelling their focus to system objectives
(gg) Planning	Study and identify the effective results that make up the objective, and activate means and methods to achieve them
(hh) Result orientation	Focus on outcome rather than process used to produce a product or to deliver a service
(ii) Personal branding	Questionnaires aimed at determining the subject's know-how related to Soft Skills

where XQ is the ability in communicating difficult content in a soft way (important part of the soft skills) and QS means the quotient of success (a complete personality, with great Intellectual Fusion, can reach a high possibility of success).

https://www.arealme.com/left-right-brain/it/.

This link will allow you to evaluate your prevailing percentage of the right or left hemispheres.

Table 1.3 Results of a survey on the importance of soft skills attributed by a sample of students

Soft skill	[%] not important	[%] not very important	[%] some-what important	[%] very important	[%] extremely important
Integrity	–	–	–	7.0	93.0
Communication	–	–	–	8.8	91.2
Courtesy	–	–	3.5	12.3	84.2
Responsibility	–	–	8.8	19.3	71.9
Interpersonal skills	–	–	15.8	22.8	61.4
Professionalism	–	–	12.3	40.4	47.4
Positive attitude	–	–	10.5	43.9	45.6
Teamwork skills	–	1.8	28.1	26.3	43.9
Flexibility	–	1.8	21.1	35.1	42.1
Work's ethic	–		24.6	38.6	36.8

The soft skills, therefore, in addition to the hard ones, harmoniously complete the personality of the engineer, like any other person, who thus has the opportunity to become a "difference maker". The importance of soft skills is confirmed by Table 1.3 [3] relating to a survey carried out on a sample of students to whom to deliver soft skill training.

1.4   The Soft Skills in Engineering and Engineering Education

1.4.1   Phases of Engineering

Engineering is a very ancient activity: it is possible to say that the most ancient profession is that of the engineer! It is necessary to observe that, since prehistoric times, human beings have made objects of use for all everyday needs.

The first scientific observations and speculations began in Ancient Times, but technicians made their objects without regard to the first scientific results, in other words, science and technique were running parallel, without, or with very few, reciprocal contacts.

Only during the Renaissance (15th–16th Centuries) did technical realizations begin to use scientific results: technology was born and this situation corresponds to the first step of modern technical development.

The upgrading of Engineering was very quick: the great development of the 17th and 18th Centuries forms the background of the modern development that we still see today.

It is useful to recognize the following phases of Engineering development:

(a) **Engineering 1.0** (up to 1970), characterized by Mechanics and Electricity.
(b) **Engineering 2.0** (1970–1995), characterized by Electronics, Automation and Informatics.
(c) **Engineering 3.0** (1995–2015), characterized by integration of Electronics, Automation and Informatics.
(d) **Engineering 4.0** (since 2015), characterized by a great development of Electronics, Automation and Informatics, by innovative disciplines, such as, e.g., integration of different scientific-technical fields, virtual and augmented reality, nanotechnology, biotechnology, biomimetic, additive manufacturing, design science and, among many others, soft skills. The Soft Skills quickly became an important part of the above-mentioned Engineering phase and are the main purpose of this book.

1.4.2   The Soft Skills as Part of Engineering 4.0

Engineering 3.0 required great hard competencies: technical schools and Universities focussed on giving students the above-mentioned competencies and graduate engineers in general had a very good education for performing the required activities, but, in general, their preparation, from the point of view of the soft skills, was not very good.

In the Engineering 3.0 period the requirements of "soft" competencies began: e.g. a teamworking relationship began to have more and more importance. In the same way, e.g., the communication requirements became ever more important.

Therefore, soft skills attained the role of being an important part of Engineering 4.0.

1.5 Aims and Contents of the Book

The aim of this book, starting from the considerations above, is to present and highlight the fundamental role of soft skills in modern Engineering.

In addition, another aim of the book is to provide some input and suggestions for a deeper knowledge of soft skills.

Finally, for the most important soft skills cited, the book hopes to provide input and suggestions on applying such skills in Engineering activities and, why not, in everyday life.

For more in-depth information about soft skills, see [5–10].

With regard to the configuration of the chapters in the book, the authors adopt the following:

(a) First, the Soft Skills considered as the most general, i.e.:
  (a.1) Communication
  (a.2) Courtesy
  (a.3) Flexibility
  (a.4) Ethics and Integrity.
(b) Second, some specific Soft Skills, related to relationships with others and with himself, i.e.:
  (b.1) With others: Interpersonal relationship, Positivity, Professionalism, Responsibility, Teamwork, Empaty, Engagement, Leadership.
  (b.2) With himself: Learning to learn, Problem solving, Digital thinking, Planning, Result orientation, Awareness, Personal branding.

(c) Two Soft Skills relative to some specific aspects of Engineering:

    (c.1) Imagination and creativity

    (c.2) Human aspects of technology

(d) Finally, the summary and concluding Soft Skills, i.e. Talent development.

(e) As conclusion, starting from above mentioned consideration, the Proposals of the authors, with the aim to upgrading the Engineering Education, from the point of view of Soft Skills.

References

1. E. da Rotterdam, *Moriae Encomium* (original edition in Latin), p. 1511
2. B. Pascal, *Pensieri e altri scritti* (Oscar Classici, 2018)
3. M.M. Robles, Executive perceptions of the top 10 soft skills needed in today's workplace. Bus. Commun. Q. **75**(4), 453–46
4. E. Rovida, G. Zafferri, Proposal about the introduction of the soft skills in the teaching of product development, in *JCM 2018 (International Joint Conference on Mechanics, Design Engineering and Advanced Manufacturing)*, Cartagena, Spain 20th–22th June 2018
5. https://www.kobo.com/it/it/ebook/the-strategy-pathfinder. Last visit 25 April 2020
6. W. Bennıs, *The Habits of Highly Effective People* (Simons and Schuster, 1989)
7. N. Negroponte, *Being digital* (Vintage book, New York, 1996)
8. D. Goleman, *Primal leadership. Learning to Lead with Emotional Intelligence* (Harvard Business Review Press, 2004)
9. K. Robinson, *The Element. How Finding Your Passion Changes Everything* (Penguin Group, 2009)
10. S.R. Covey, *The Speed of Trust* (Simon and Schuster, 2006)

Communication

2

How building or own behaviour.

2.1 Introduction

One statement is particularly significant to discuss the topic:

> To understand himself, man needs to be understood; to be understood, he must understand [1, 2].

Human beings, all of them, interact by communicating. The word *communicate* comes from the Latin *Communis agere*, or *Put in common.*

Indeed communicating means transferring a message in a way that allows the person receiving it to interpret it, attributing the same meaning the message has for the sender. This aspect presents quite a problem, if we consider that communication features three basic axioms [3]:

1. There is no non-behaviour, as behaviour is, in fact, the basis of human relationships and, therefore, it is, itself, the basis of interpersonal knowledge
2. Non-communication is impossible, as every behaviour is, in fact, a communicative message

© Springer Nature Switzerland AG 2022
E. Rovida and G. Zafferri, *The Importance of Soft Skills in Engineering and Engineering Education*,
https://doi.org/10.1007/978-3-030-77249-9_2

3. It is impossible to avoid influencing others, as every form of communication has an effect on the individual's perception of reality.

One must keep in mind that, what is important in a communication process is not what we say but how we say it, how we are listened to, how we are understood, how we are interpreted, how we are remembered and, finally, how our interlocutor reacts to this.

The communication process is like an *iceberg* (Fig. 2.1): we are accustomed to only seeing the part above the water line, and we either do not see, or underestimate, the submerged part, which is actually more important and significant.

The apex (or referring to the image, the tip of the iceberg) constitutes verbal and written communication. The "submerged" part represents non-verbal communication, in other words facial expressions ("face"), body movements, clothing, proxemics, the

Fig. 2.1 The communication iceberg

THE ELEMENTS OF COMMUNICATION

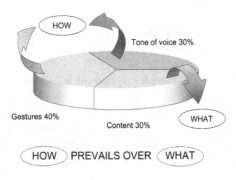

HOW PREVAILS OVER WHAT

Fig. 2.2 The "how" of communication takes precedence over the "what"

ability to listen and neuro-linguistic programming (NLP) (see Paragraph 2.8). Non-verbal communication is a predominant part of communication.

We must also observe how the combination of gestures and tone of voice makes up 70% of communication. In other words, the "how" of communication over the "what" (Fig. 2.2).

2.2 The Basic Rules of Communication

"Personality" plays an important role in communication: it is the result of several factors, some of the most important of which are listed below:

- Personal history
- Behaviours
- Motivation systems
- Feelings
- Culture
- Context
- Status
- Psychosocial roles

- Values
- The Unconscious.

It must then be said that the process of understanding the meaning of communication unfolds through the following factors:

(a) The filter, or the complex system of values that belongs to each individual; subsequently, each person can choose certain elements of communication and discard others
(b) The halo effect, related to the fact that a word or an idea can trigger personal associations and can, therefore, either hinder or help communication, depending on the links triggered.

Figure 2.3 represents information between two people. It is interesting to observe the following. If the theoretical content of communication is 100, then what is effectively said is 70, while we receive 40, understand 20 and remember 10. Hence, every communication presents a strong deterioration between what the person communicating wants to communicate and what the receiver later remembers.

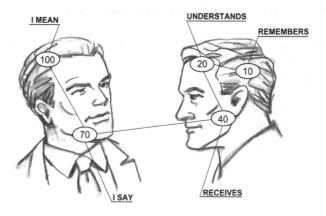

Fig. 2.3 "Deterioration" of communication"

The basic elements of communication are listed below:

(a) **Source**: the person communicating
(b) **Recipient**: the person receiving the communication
(c) **Message**: the content of communication
(d) **Coding**: the link between the content intended by the source and the system of signs used (e.g., written or spoken words and images)
(e) **Channel**: the path (e.g., visual or auditory) through which communication happens
(f) **Decoding**: the reverse of coding, in other words, the recipient's perception of the sign system and the reconstruction of the content intended by the source (obviously, minus the deterioration exemplified in Fig. 2.3)
(g) **Feedback**: the recipient's "response" to what he understands of the communication (it is basically the verification of communication)
(h) **Context**: everything that surrounds communication.

Figure 2.4 exemplifies the above.

The source structures the content he wants to convey and "assigns" it to a system of signs (written words, spoken words or images) transferred through a physical medium.

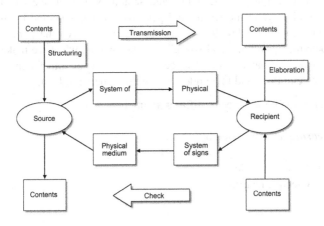

Fig. 2.4 Logical diagram of communication

The recipient receives the system of signs through the physical medium and reconstructs the content, obviously keeping in mind the "deterioration" exemplified in Fig. 2.3.

During check, which is the feedback of communication, the recipient transfers the content to the source through a system of signs and a physical medium.

For further information see [4–10].

2.3   Components of Communication

The main components of communication are generally the following:

(a) Content that is communicated
(b) The context in which the communication takes place
(c) The relationship that develops between the interlocutors.

The ability to listen is extremely important. It is essential to listen in order to communicate well.

But this also depends on how we listen, as listening can be: *passive, attentive, active or empathic.* It is passive when we only hear the sound of the voice. It is attentive when we grasp the rational meaning of the words and sentences. It is active when we look beyond the words and perceive what lies behind them (voice-tone-attitude). It is empathic when we give the person speaking "psychological space", in other words, when we make him feel he has an open "space" to delve deeper into feelings, show emotions and feel understood without being judged.

Hence the importance of differentiating between:

Hearing, characterised by:

(a) Passive attitude
(b) Selective hearing

(c) Rapid elimination of what is heard
(d) Lack of interest
(e) Lagging conversation
(f) Various noises
(g) Anonymity.

Listening, characterised by:

(a) Focusing
(b) Absorbing
(c) Interpreting
(d) Being involved
(e) Understanding and retaining
(f) Forming an opinion
(g) Attributing emotions to the content.

In other words, knowing how to listen is an art. Some guidelines to best develop this art are the following:

- Do not be hasty in drawing conclusions. Conclusions are the most ephemeral part of communication
- What we see depends on our own point of view. In order to see correctly, we need to change our point of view
- If we want to understand what someone else is saying, we need to assume that they are right and ask them to help us see things and events from their perspective
- Emotions are fundamental cognitive tools, if we know how to understand their language. They do not tell us about *what* we see, but of *how* we see: their code is relational and analogical
- A good listener welcomes the paradoxes of thought and communication. He considers disagreements as opportunities to practice in a field he enjoys: he knows how to creatively deal with conflict
- Moreover, a certain sense of humour is not a bad thing when listening.

2.4   Factors that Impair Effective Communication

In terms of deterioration in communication (Fig. 2.3), factors that can be an obstacle to good communication are grouped by whether they are generated by the sender or by the receiver:

(a) **Source**:
- Stress and worry
- Emotional block
- Hostility
- Distraction
- Memory of past experiences
- Stereotypes
- Physical environment
- Tone of voice
- Mental noise
- Distractions
- Hasty communication
- Length of message
- Use of slang terms.

(b) **Recipient**:
- Wrong interpretation
- Lack of understanding
- Excessive effort in other activities
- Differing meanings attributed
- Boredom.

For further information see [11–14].

2.5   The Question Game

The essence of good communication lies in talking with someone and not in talking at someone. And the best way to talk "with someone" is to ask questions.

Neutral questions that are not aimed at producing any specific emotional response can be divided into closed questions, which elicit only two types of answers: YES or NO. We, therefore, use closed questions to clarify, to affirm, to summarise and to confirm.

Open questions start with: Who, How, What, Where, When, Why. We use them to gather information, invite dialogue, engage both parties, start a relationship, and to better understand.

The advantage of knowing how to ask questions can be traced to the following: we obtain and give information, we develop relationships that relax and boost confidence, we give value to the interlocutor, we gain control over the situation, and we obtain greater detail.

To show that we have carefully payed attention and to give the interlocutor a sign of motivation, it can be useful to use some connecting expressions before moving on to questions (i.e., of course, very well, I am sorry to hear this, let's go over the main topics of what you are saying...).

It is also essential to know how to reword, or concisely repeat what the person has said, in order to allow us to respond to the topic by showing attention, respect and interest for what our interlocutor is saying:

- If I am not mistaken, you're telling me that...
- So according to you...
- So recapping....
- You're saying that...

2.6 Communication Channels

Communication channels are summarised in Table 2.1.

It is also important to identify the efficiency and effectiveness of communication (Table 2.2). They are strictly connected to the same channels.

The "**Verbal**" channel includes the content and the structure of discourse. The *content* relates to the informative aspect, which can reveal either facts or subjective elements, show constructive, destructive or blackmailing traits, be vague or precise. The

Table 2.1 Communication channels

Verbal	(a) Be intelligible: speak the same language (b) Ask stimulating questions: to encourage reflection and understanding (c) Be impactful: to prevent others from forgetting
Tonal	(a) Vary the pitch: to increase persuasion (b) Know how to pause: to show confidence (c) Underline: to highlight a specific, important item
Non-verbal: body	(a) Body language: to show mastery (b) Presence: appearance is what makes the first impact (c) Self-projection: to give impetus
Non-verbal: gestures	(a) Controlling gestures: gestures help us communicate (b) Facial expressions: to strengthen the message (c) Visual contact: seeing means involving

Table 2.2 Efficiency and effectiveness of communication

Efficiency	Effectiveness
Presence	Have a clear idea of what you want to say
Initial silence	Adapt your own style to that of the interlocutor
Initial question	Do not speak quickly (pauses help both the speaker and the listener)
Use silence	Do not be verbose
Gestures	Try to be practical
Eye contact	Change the tone (do not be monotonous)
Emphasis	Show that you value the interlocutor by referring to his knowledge
"Us" form	Look into the eyes

structure is determined by the subject's intentions towards the interlocutor: issue orders or make requests, investigate or walk away. The skill to continuously scan the dialogue can also be included in this area: respond in kind, interrupt or let the other person talk.

The "**Tonal** channel" includes the tone and rhythm with which the verbal elements are emitted.

The *tone* refers to the pitch and intensity of the voice. Thus a person can talk in a high or low-pitch, determined or undefined way that can either be devoid of variations or emphatic. *Rhythm* concerns speed and frequency: it is possible to speak quickly or slowly, fluently or in bursts. To this we can add the *sound expression*, which involves all further sounds apart from words, such as sighs and coughs.

The "**Non-verbal**" channel, commonly defined as "body and gestures language" (body movements and gestures), includes the areas described below: gestures, posture and proxemics.

Mimicry: the combination of facial expressions, movements of the head and manifestations, such as turning pale or blushing. The gaze, a powerful communicative skill, also belongs to this category. *Gesticulation*: the movement of arms and hands. *Posture*: the position adopted by the body, which can be leaning forward or backwards, standing upright or curled up. It includes the relationship with the position of the interlocutor: facing, sideways or diagonally. *Proxemics* refer to space and distance as a factor of communication.

For further information see [15–17].

Figures 2.5, 2.6, 2.7, 2.8, 2.9, 2.10 and 2.11 show examples of non-verbal communication.

2.7   Attitudes

Attitudes can be defined as mental states, which have been created over time and which dynamically influence the individual, preparing them to react in a specific way to certain situations. They manifest not only in verbal communications but in all expressive situations, and underpin our way of relating to the environment.

From the perspective of interpersonal relationships, attitudes are permanent predispositions to act in a certain way.

Sadness and pain   Disgust   Rage

Sadness   Fear   Despair

Fig. 2.5 Examples of facial expressions

Attitude is the key to success. This expresses the mental condition concerning what happens, or simply the way of seeing things.

Some important items concerning attitudes are listed below:

- The mental attitude affects our behaviour: it is not always possible to mask our feelings towards the interlocutor
- Attitude determines the level of job satisfaction
- Attitude influences anyone coming into contact with us
- Attitude is not solely reflected in the tone of voice but also in the physical posture of the body, in the facial expressions and in other types of meta-language
- An attitude is never static or unchangeable: it is up to us to decide the most appropriate attitude for the situation

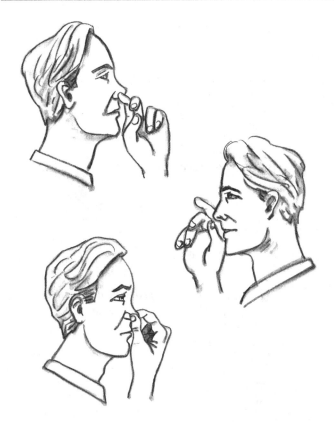

Fig. 2.6 Examples of stress

- It is important to start the day by thinking of the positive aspects of the job, connection, meeting and so on.

In unpleasant situations, we should focus on the positive side of things and on the activities that allow for an appropriate sense of proportion. If at all possible, it would be appropriate to avoid quarrel some people and conflict situations. It is also important to share our positive feelings with others when things are going well: the attitude is "contagious".

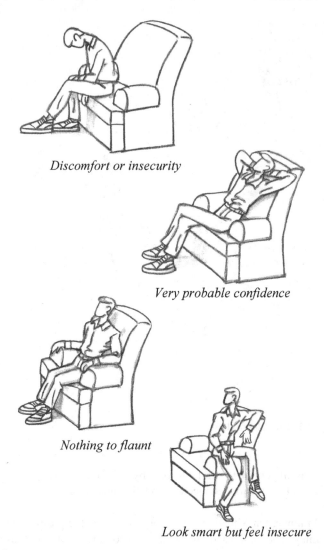

Discomfort or insecurity

Very probable confidence

Nothing to flaunt

Look smart but feel insecure

Fig. 2.7 Examples of postures

The relationship is born within the communication process, which first of all means reading the interpersonal dynamics at hand and adopting a style of communication that is consistent

Fig. 2.8 The crossing of legs involves many variations, which have as, a common element, the closure towards the outside

with the environment, possibly oriented to the interlocutor (empathy) and aimed towards the objective of the communication itself.

The main communication factors—from which the behaviour depends—have different ratios as indicated in Fig. 2.2.

Fig. 2.9 Examples of handshake (social zone)

Hence the need to choose the appropriate communication channels. The following steps are interesting:

- Step 1 is authenticity: understanding who we are, showing this to other people, which turns into communicating our own values to make them understandable, sharing them and using them to guide ourselves (such as, for example, internal clarity, sharing, motivation, drive to take action, constant reinforcement of intangible assets, the basis for evolution and development)
- Step 2 is clarity: the message must be simple
- Step 3 is continuity: communication must not be interrupted
- Step 4 is credibility: communication must occur in a climate of trust as the source must be acknowledged as valid and reliable.

Still looking good   Not buying what you're selling

Hostile attitude

Fig. 2.10 Examples of crossing arms

Communication rationale can be derived from the level of fusion between the "*facts*" dimension and the "*personal opinions*" and "*feelings*" dimensions. Therefore a message should be structured as follows:

F (facts) O (opinions) E (emotions) = FOE

Identifies *descriptive* communication, where the main principle is not to pollute facts with interpretation data; it responds

Uncertainty   *Positive expectancy*   *Telling a lie*

Fig. 2.11 Examples of gestural communication

to the need to place the interlocutor before a neutral message, which we know will be interpreted. It is "**Rational
Communication**" oriented towards a solution to problems and
group decision-making.

O (opinions) F (facts) E (emotions) = OFE

It identifies generally **coercitive** communication that, taking
advantage of sentimental aspects, places the interlocutor before
interpretational data.

E (emotions) O (opinions) F (facts)   = EOF

It identifies basically **manipulative** communication aimed at
influencing the interlocutor on an emotional level, playing with
the more essentially relational and empathic aspects.

2.8   Neuro-Linguistic Programming

Neuro-linguistic programming (NLP) is based on certain specific
topics, which can be attributed to the following factors.

Neuro: refers to our nervous system. The crossroads of our five
senses: sight, hearing, touch, taste and smell.

Linguistic: refers to our ability to use language, and how our
words and sentences reflect our mental world. Moreover, it

also refers to our secret language of posture/gestures and habits, which reveal our thoughts, our beliefs or, in one word, our "Self".

Program: our way of thinking, our feelings, and our actions are habitual programs we can change by improving our "mental software".

Of the 5 senses, those most in use in NLP communication are:

- Sight (V)
- Hearing (A)
- Touch (K).

Ultimately, human beings can be divided into three categories:

- **"Visual"**, if they are mainly sensitive to shapes, colours, and movements
- **"Auditory"**, if they mainly pay attention to sounds, noise and word
- **"Kinaesthetic"** if they are predisposed to grasping taste, smell and touch-related aspects of their environment.

Each of us has a more developed sense, compared to others, which we use to represent reality.

According to the above, our reactions lead to facial expressions that, if correctly interpreted, can change our interlocutor's reaction.

The category we belong to can be recognised by the words and the gestures we most commonly use. In fact, each of us tends to preferably use and perceive some words, compared to others. Regarding the senses, smell and taste are, for example, less developed among Westerners, compared to other populations.

Hence, attention should not only be placed on the words of our interlocutor, but also on what their body is subconsciously telling us. With a little attention and practice, they will reveal the interlocutor's sensory channel.

If, for example, they look up while we are talking, they are telling us that they are "visually" representing what we are saying. If, on the other hand, their gaze moves downwards to the right, if they sigh deeply, if they caress the desk top or touch or weigh up the objects around them, their body is telling us that their most predominant sense is "kinaesthetic". If, however, every now and then they interrupt themselves and listen to sounds or noises coming from the next room, or if their gaze moves horizontally to the right and to the left, if, when they are listening to us, they perch their head on the palm of their hand with their head slightly tilted, as if they were making a call, or they stroke their chin or touch their nose or ears or areas of the mouth, they are telling us that they are on the "auditory" channel.

If the sense of "sight" is prevalent in our interlocutor, we can invite them to clarify their thoughts: let us show them our point of view with words evoking colours, nuances and movement. We can show them documents, graphs, photographs and images to support what we are saying, so that they can form a picture of the situation that is as clear as possible. If they are mostly "auditory", we can ask them to listen carefully to what we are saying, in order to better and more quickly harmonise our objectives and our actions. Finally, if they are mainly "kinaesthetic", we can try to shake them up, having them touch the advantages of our proposal, something which will make them more confident and more at peace in future situations.

Recognising and discovering to which of these three classes our interlocutor belongs is very important, and it allows us to choose words, adjectives, verbs, adverbs which—case by case—will make our messages come across more clearly and directly.'

If, independently from the prevalent perceptive channel, we switch on all our senses, our messages will be better received and remembered for a longer time. Therefore, let us learn to refine our perceptive abilities.

For further information see [11, 12, 18].

2.9   Intercultural Communication

The cultural background of an individual is very important for his communication. Indeed, a very important element is the cultural profile of a person. It is, essentially, constituted by family, religion, education, culture, profession, social class, gender, ethnic background, generation, neighbouring countries and social relationships.

It is crucial to identify the dominant orientations in a culture and, at the same time, to pay attention to the variations and the changes, which characterise the individual.

Within a culture, there are various factors that influence groups and individuals, among which we can mention the generation (for example, Baby Boomers (1946–1964), Generation X (1965–1980), Generation Y (Millennials (1980–2000)), Generation Z (2000–2010, Generation Alpha (2010–2020), Generation C (it's a conceptual attitude versus connectivity independent passage).

Stereotypes and prejudice are reference points that are necessary for the perception of others. It is then up to us to overcome them, judging the people in front of us objectively.

Every culture is oriented and has specific preferences such as, for example:

(1) relationship with the environment (control/harmony), (2) relationship with time (polychronic/monochronic, rigorous/vague, past/present/future), (3) relationship with oneself (being/acting), (4) style of communication (very contextual/not very contextual, direct/indirect, expressive/instrumental, formal/informal), (5) relationship with space (private/public), (6) relationship with power (hierarchy/equality), (7) relationship with individual-group (individualism/collectivism, universalism/particularism), (8) relationship with others (competition/collaboration), (9) need for structure (order/flexibility), (10) way of thinking (deductive/inductive linear/systemic).

It must also be kept in mind that cultures change and evolve based on events and on the various changes in the macro-environment. The main factors, which determine them, can be mostly traced back to the following (E.S.T.E.M.P.L.E.):

- Economic
- Social
- Technological
- Ecological
- Media
- Political
- Legal.
- Ethical.

The mechanism of "intercultural" perception is also important, for example:

Ethnocentrism: we think we are at the centre of the world

Resemblance: we think others think the way we do

Interference: we only see what we want to see

Projection: we project our unmet desires onto others.

Cultural difference factors are also essential in communication/negotiation, such as for example:

1. The way you lead relationships
2. Assessment of time
3. The form of communication
4. The weight of hierarchy
5. The consideration of status
6. Group dependency
7. Acceptance of differences
8. Adaptability to change.

What is considered a correct behaviour in our country might be seen as the contrary in others. So, besides learning the rules of business itself (*Hard Skills*), we also need to know, on top of language, the culture (*Soft Skills*). We must, for example, consider the structure and the meaning of hierarchy. A strongly hierarchical culture tends to put "everyone in their place", and considers "changes of status" as something that will happen rarely and slowly. Addressing a subordinate by skipping the direct superior is greatly offensive.

The time factor is also very important. For example, in Western cultures it is generally considered a precious resource, which must be used sparingly, while in other cultures (Far and Middle East) it is, at times, a more elastic concept, which is less important, compared to other elements of social coexistence.

To this end, it is interesting to see how time is considered in various cultures (Table 2.3).

Table 2.3 The concept of time in various cultures

Polichronic (Common in Latin countries, in Southern Europe, in the Mediterranean and in Japan)	*Monochronic* (Popular in Germany, Northern Europe and North America)
Multiple activities are carried out at the same time	One thing at a time
Interruptions and changes of activities are accepted	We devote ourselves totally and exclusively to the task at hand
We communicate by referring often and extensively to the context	We communicate without referring to context or not often
Relationships between individuals are more important than achieving the result	The execution of the project or of the task at hand has a higher priority compared to relationships with other individuals
Plans and projects are easily and often modified	An established programme is followed scrupulously
Priority is given to people who are emotionally close to us	An effort is made not to disturb other people. Confidentiality and discretion are the rule
Exchanges and lending of personal objects are common and well accepted	Property is well defined. We lend something only if we are forced to do so
Punctuality is quite relative	Excessive punctuality

(continued)

Table 2.3 (continued)

Polichronic (Common in Latin countries, in Southern Europe, in the Mediterranean and in Japan)	Monochronic (Popular in Germany, Northern Europe and North America)
Relationships are stronger and more lasting	Relationships are more superficial and short-lived
People are impatient and tend to move straight to action	People are slower, more methodical and less involved
Binding commitments worry people	The most binding commitments are the ones based on time, date and duration

For further information see [1, 2, 19].

We can therefore state that in general "hard skill" don't change very often, while "soft skill" may change continuously.

References

1. T. Hora, *Self-Trascendence* (The PAGL Foundation Inc., 2008)
2. T. Hora, *Encounter with Wishdom* (Bookone, 2018)
3. Watzlawic, P. Beavin, J. Helnik, D.D. Jakson, *Pragmatica della comunicazione umana* (Astrolabio, 1978)
4. C. Ruggiero, M. Luccio, *Studiare la comunicazione* (Maggioli, 2013)
5. R. Heller, *Communicating clearly* (DK Pub, 1998)
6. S.M. Markle, G*ood frames and bad. A grammar of frame writing* (J.Wiley & Son, 1964)
7. B.F. Skinner, *The science of learning and the art of teaching* (Cambridge Mass, USA, 1954)
8. J.J. Guilbert, *Taxonomy of intellectual processes,* Didakta Medica 2 (1971)
9. B.S. Bloom, .*Taxonomy of educational objectives* (Handbook I.D.McKay, New York, 1956)
10. M.A. Bierstadt, *Mapping the phenostructures of didactic sequences* (School of Education Malmo, 1964)
11. P. Watzlawick, *Istruzioni per rendersi infelici* (Feltrinelli, 2013)
12. P. Watzlawick, J.H. Weakland, R. Fish, *Change* (Astrolabio, 1978b)
13. D. Chieffi, *Comunicare digitale* (Centro di documentazione giornalistica, 2018)

14. S. Bentivegna, G. Boccia Artieri, *La teoria della comunicazione di massa e la sfida digitale* (Laterza, 2019)
15. A. Lowen, *Il linguaggio del corpo* (Feltrinelli, 2013)
16. E. Berne, *A che gioco giochiamo* (Bompiani, 1964)
17. W. Allen, *A proposito di niente* (La nave di Teseo, 2020)
18. A. Schopenauer, *L'arte di ottenere ragione* (Adelphi, 1991)
19. G. Bateson, *Verso un'ecologia della mente* (Adelphi, 1977)

Courtesy

3

How to win the respect of others.

3.1 General Principles

The word "courtesy" has the etymology of "Court": the courteous person who was used to living at Court, following the habits of people of high lineage, as opposed to "villano", who was the peasant. This citation is significant: [1] "Cortesia e onestade è tuttuno: e però che nelle corti anticamente le virtudi e li belli costumi si usavano, siccome oggi si usa tutto lo contrario, si tolse questo vocabulo dalle corti e fu tanto a dire cortesia quanto uso di corte". (*Courtesy and honesty are one: the fact is that virtues and beautiful costumes were used in ancient times; since the opposite is true today, this word has been removed from Courts and it is the same to say courtesy or a Court custom*).

It could almost be said that, even in Dante's time, there was "nothing new under the sun!"

However, today the word means behaving kindly, affably and correctly towards others.

It is important not to confuse courtesy with foolish weakness towards our interlocutor: respect, which generates courtesy, is the balance between the fear of one's neighbour, which leads to uncritically accepting everything others either do or say, and

© Springer Nature Switzerland AG 2022
E. Rovida and G. Zafferri, *The Importance of Soft Skills in Engineering and Engineering Education*,
https://doi.org/10.1007/978-3-030-77249-9_3

arrogance, which leads to prevaricate, systematically contesting every statement of the other. If you want to improve the world, the first step is to improve yourself, and courtesy plays an important role in the process.

In this regard, a major obstacle is the excess of ego. This excess can manifest itself in various ways, including:

(a) Try to highlight your brilliance, true or presumed, in every way
(b) Continually seek the approval of others
(c) Always be defensive.

Knowledge, especially of yourself, is important to get rid of the ego. Albert Einstein said:

> more the knowledge, lesser the ego
> lesser the knowledge, more the ego.

The ego often presents itself as self-confidence, but it is soon recognized as arrogance and, in the end, as insecurity.

The ego amplifies information beyond its meaning, selects only what makes our thinking valid, tends to discredit information that contradicts our stand, manipulates information aimed at validating our ideas, and "manufactures" situations, information that never existed.

Some ideas for the "cure" of the uncontrolled ego can be traced back to the following:

(a) Humility, accepting the principle of ignorance
(b) Curiosity, admitting that you do not know
(c) Sincerity, recognizing the value of others
(d) Listening skills, attention to others.

3.2   Courtesy in Daily Life

As mentioned in the previous paragraph, courtesy is generated by the respect each of us must have for others. Hence, courtesy is respecting others, in other words, not doing to others what we would not like were done to us.

In everyday life, courtesy can be attributed to the following consideration.

Each of us, in his life, comes in contact with many people every day: it would be important to think that each of our gestures, each of our words, leaves an impression on others, either large or small. Hence, the courteous attitude "sows positivity". Furthermore, it does not hurt to think about mirror neuron theory: our attitudes, in many cases, induce a sort of imitation in others. Our courteous attitude can, therefore, have a beneficial spreading effect like wildfire.

The courteous person knows how to be empathic; he steps into others' shoes, understanding their problems and needs.

3.3 Courtesy in Engineering

The engineer is a person who almost always works in contact with other people: customers, superiors, colleagues and subordinates.

Hence, courtesy is particularly important. The way one behaves towards others is, in fact, significant to create a harmonious work environment.

Only too often do we see work environments in which the manager treats colleagues and subordinates with arrogance (if not with downright rudeness). Such an environment is, of course, negative in terms of efficiency and productivity. The considerations made in paragraph 3.2, therefore, also adapt very well to engineering and, ultimately, to all professions. Courtesy in the professional field costs only a little attention and goodwill, but can yield a lot in terms of work results.

3.4 Courtesy in Teaching

A valid way of "teaching" courtesy, based on the experience of the authors, may consist in assigning a group, preferably a small one (15–20 people), to carry out a project, allocating a certain time for development. After the time has elapsed, the teacher performs a public discussion, collectively asking for feedback and comments on each possible individual errors. For example, after the exercise in Fig. 3.1 has been assigned, each student develops a concept and a proportionate system construction solution within the time allocated to perform the required task. The resulting system, which is easy to recognize, can be traced back to a carpenter's wise.

Courtesy can be "taught" directly by example: the teacher trains students to conduct a "polite" discussion, for example:

(a) Not interrupting the speaker and asking to speak before speaking
(b) Making "constructive" and "non-destructive" comments on the work carried out
(c) Avoiding offensive judgments of mistakes made by others
(d) Objectively examining the pros and cons of each work.

By "containing" the arrogant, encouraging the timid and highlighting the general advantages of such an attitude, we can achieve overall enrichment of the group.

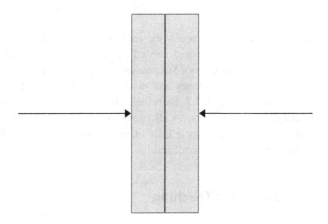

Fig. 3.1 Example of exercise assigned in the "Design Methods" course; it is a question of representing a mechanism to exert two opposite and equal forces to compress the two parts to be glued. The collegial discussion of the work, if correctly conducted, is an excellent exercise of respect and courtesy

As further example, it is possible to consider the valve represented in the Fig. 3.2. The students are invited to draw the assembly of the valve in orthographic projections and to propose new constructive solutions, e.g. for the connections, for the control of the ball and for the manufacturing processes utilized to realize the different parts of the ball. Also in this case, the orderly discussion is a good exercise of courtesy and respect of the students between each other and with the professor.

The examples presented in the Figs. 3.1 and 3.2 was utilized, as said before, by one of the authors in the exercises of Machine Design and was very good applications of brainstorming. This creative technique has the aim to develop new ideas from a discussion among all the participants of the group. The brainstorming is a technique often used in the teamwork (see also such Chapter in the present book) and is very important and useful from the didactic and technical points of view. The opinion of the authors is also that brainstorming, if organized and conducted correctly, is also a very good exercise of courtesy and reciprocal respect: e.g. by speaking without interrupting others and constructively discussing their opinions. Moreover, these

Fig. 3.2 Example of assembly drawing to be studied by the students. Courtesy of Technical Drawing and Machine Design of the Mechanical Department of Politecnico di Milano

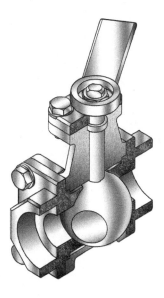

concepts often recur in the various Soft Skills deal with in the book, albeit examined from different points of view.

Reference

1. D. Alighieri, *Il Convito* (Tipografia Camerale, 1831)

Flexibility

4

Ability to adapt to change (i.e. resilience).

4.1 General Principles

Adaptability to change, or flexibility, can be defined as the ability to perceive, in a timely manner, changes in a situation, in a context, in objectives and in moods, in order to rapidly reassess matters and alter the action to be implemented. Flexibility is, therefore, the ability to keep in step with the times for a better future.

We often hear the word "resilience", understood as an individual's ability to react to unexpected and sudden situations that are, in any case, objectively considered negative.

In this regard, a famous joke by Woody Allen says that the maturity of a person is measured by how they react on waking up in the morning… in their underwear, in the heart of the city! It is, of course, a paradoxical joke but, as is often the case with paradox, it is based on some amount of truth.

It is important to notice how the subject's reaction, and consequently their behaviour, can (and must) vary in relation to the entity of the change. If this is fairly small, the subject's reaction can be traced to a variation in attitude, born in the mind.

© Springer Nature Switzerland AG 2022

E. Rovida and G. Zafferri, *The Importance of Soft Skills in Engineering and Engineering Education*,
https://doi.org/10.1007/978-3-030-77249-9_4

If, however, the change is great, the subject's reaction must come from a variation in paradigm, in other words, in the way they perceive reality.

For example, a delayed train (a minor change) can create momentary irritation that, however, can simply mean sitting in a corner and reading a newspaper (variation in attitude).

An individual who peacefully reads an interesting book in a corner of a public area is quite another matter. At a certain point, a father walks in with three noisy restless children (a change that is not so minor), whom he ignores. The subject's initial irritation dissipates immediately, turning, on the contrary, into solidarity towards the father, as soon as he finds out that the father was coming from the hospital where the wife was being treated for a serious illness and dead (change of paradigm).

If only small changes are desired, then we need to change our behaviours, while if, on the other hand, greater change is the objective, then we must change the paradigms.

The current situation is complex and dynamic, characterised by quick and often unpredictable changes. What is valuable today might be valued less in six months, and maybe even less in a year. This consideration further confirms the need to quickly adapt as situations change. The renewal must be carried out in a harmonious way within the four directions that characterise a complete person, precisely: *physical*, *social/emotional*, *mental* and *spiritual*.

An unfortunately common mentality, even today, opposes change. "Why change? We have always done things this way." How often have we heard this phrase repeated to this day, if not the even worse statement, "Everyone does things this way!". Repetition and plagiarism, which are still only too common, hinder a rational and innovative approach. Subsequently, you soon end up being "left behind".

We can also observe, through biomimetic comparison, that animals living a better and longer life are neither the strongest nor the most intelligent, but those which can better react to change, adapting more quickly to metamorphosing situations (Charles Darwin 1809–1882).

We must, however, be wary of the opposite situation: change for the sake of change, often without realising why. After a decision is made, we often hear the words "perhaps the other solution would have been better", as the pros and the cons were probably not thoroughly assessed (or not assessed at all!).

Flexibility does not mean constantly changing opinion. It is safe to say that among our acquaintances there are examples of people who take a path and then, before reaching the end of it, shift to another and yet another, which they never complete. This is a "very effective" way to collect disappointments and frustrations, and to waste both time and money.

There are, therefore, two types of errors, which relate to two types of people: those who resist change in favour of "nostalgia" ("we have always done things this way"), and those who constantly change ("it might be better to do it this way"). The most balanced people, on the other hand, keep up with progress to create a better future.

For those who enjoy paradoxical jokes, it might be worth referring to the typically British masterpiece of paradox, which is the tale of "Alice in Wonderland" and, particularly, to the dialogue between Alice and the Cheshire Cat, portrayed by Walt Disney's drawings:

Alice: "Tell me which path I must take."
Cat: "It all depends on where you want to go."
Alice: "I just want to go somewhere."
Cat: "Well then, if any path is fine, the important thing is to go all the way."

Indeed, abandoning the chosen path, of course when there are no logical and rational reasons to do so, is a mistake that often carries a high price.

We can observe that there are two categories of "things", those which can be changed and those which cannot. Being flexible means being able to distinguish them, having the strength to change the former and to deal with the latter, trying, however, to see the positive sides.

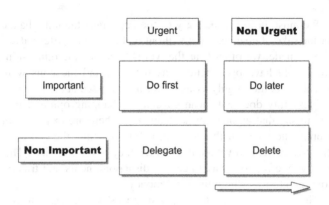

Fig. 4.1 The Eisenhower box/matrix

The Eisenhower matrix can be considered when talking about the ability to make quick decisions in relation to the variations in context. The 34th President of the United States from 1953 to 1961 reached a very high level of productivity, also by applying the Eisenhower Box [1]. This box (Fig. 4.1) encourages more rational decisions by classifying the things to do as more or less "important" and "urgent.

According to this matrix, things to do are classified based on their importance and urgency. Important and urgent decisions are, of course, the most critical, while the important but not urgent ones are particularly interesting, as they can be thought over and analysed.

Urgent decisions that are not important can be delegated, while non-urgent and non-important ones can be postponed to a later date.

4.2   Daily Life

In daily life, there are many circumstances in which situations change. Even in many trivial situations, often with no great consequence, each individual has to quickly change plans. Flexibility, therefore, comes into play frequently, even in daily life.

It is part of daily experience to see people on the verge of a crisis due to a missed appointment, a delayed train or flight, a change of schedule or any other trite circumstance.

A simple example. Just before an important dinner, the owner of a restaurant is informed that the chef will not be present that evening. Rather than panicking, the lady, a knowledgeable gourmet, responds to the challenge by cooking the meal herself: the dinner is a success.

A good method to overcome such crises with a flexible approach could be to study them from a broader perspective, almost by applying a temporal "zoom in". These setbacks and the relative crises are often forgotten after just a few hours, or we might even find out that the event had a previously unimaginable positive outcome. It seems necessary, in this case, to wonder if it was worth it.

We could mention the well-known Chinese proverb: *Before crying for a misfortune, consider if it is not a blessing.* And, conversely: *Before rejoicing for a blessing, consider if it is not a misfortune.*

4.3   Engineering

The engineering profession is very complex and presents several facets. As a rule and with considerable approximation, we can say that engineering is the profession that deals with designing, producing, distributing, using and disposing of technical products. In this sense, we can refer to the product's lifecycle. Table 4.1 shows the general and specific phases.

The project, which is the result of the designing phase, "shifts" to the production phase, whose outcome is the product. Production can be broken down into two phases: production of parts and assembly.

The product, once made, must be distributed. This operation turns it into a product in use. The distribution phase can be analysed through packing, loading, transport and unloading phases.

The product in use starts having the same function that was expressed in the project's objective. Its use involves performance

Table 4.1 The lifecycle phases of the product

General Phases	Specific phases
Production	Manufacturing of parts Assembly
Distribution	Packing Loading onto a transport vehicle Transport Unloading from a transport vehicle
Utilization	Performance of services Maintenance Ergonomics Risk prevention Aesthetics
Disposal	Liquidation Recycle

(which specifically matches the function with the objective's specifications), maintenance (which keeps the running of functions unaltered or almost so in time, at least to a certain extent), ergonomics (which aims to make the product user oriented), aesthetics (a variably important need based on the type of product), and risk prevention (the need for the product not to cause physical harm to people).

The utilization phase is limited and the product goes out of use. As such, it can either be eliminated and, thus, become waste, or be recovered and become a resource that re-enters the lifecycle of the product itself in some other form.

The resource can be a product of the same type as the initial (used) product, a lower level product, a material or a group of materials, or energy.

In all the above phases, the engineers involved must be ready to understand the needs they reveal. A designer, in particular, must be ready to reassess things based on the emerging needs, initially considering simulations carried out during the design phase, which used virtual reality and augmented reality tools. Subsequently, when the manufactured product is distributed and used, problems and consequent needs might emerge that must be taken into account for the modifications required in the following editions of the product itself. Feedback aimed to modify the product can also emerge for production, distribution, use, and why not, elimination engineers.

All engineers must, therefore, have quick reflexes to understand the needs of the market (in other words, of the actual and potential users), and to carry out the necessary corrections.

Engineers are, therefore, among the professionals who must especially be ready to adapt based on changing needs, also of the market, which can be very quick.

A very high risk in the industrial world, a risk that was mentioned in relation to the general principles, also involves the engineer's job. It consists in blocking change and innovation with the statement that is repeated only too often even today, "we have always done things this way, why change?".

Figure 4.2 is significant, as it shows the importance of companies' ability to change.

This is a double entry table, which includes four cells, each of which corresponds to a specific attitude of the company. The

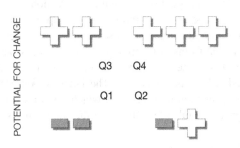

THE 4 QUADRANT

Fig. 4.2 A company's potential for change versus the current level of transformation 4.0—change in the Porter's value chain from push to pull

(Courtesy of)  Officina della Conoscenza

horizontal coordinate corresponds to the current level of transformation 4.0, while the vertical one corresponds to the potential
for change. Each "cell" corresponds to the type of company.

Q4: the potential for change is high and they are "in movement"
    in terms of current level of transformation. Hence, these
    companies have a very great chance to succeed in the evolv-
    ing market. These companies can be an example for others
    and can also be in the position to work with rivals.
Q3: the potential for change is high and theoretically "in rush",
    in terms of current level of transformation.
Q2: medium potential for change and current level of transfor-
    mation is "in movement". They are, therefore, proactive
    companies.
Q1: low potential for change and level of transformation: there-
    fore, they are at risk.

A great entrepreneur says: *If a company does not follow the
times in which it operates, it will become old and I do not want
this to happen. Always change to remain competitive.*

Staying up to date, therefore, is very important for an
engineer, as for all professionals, and in the field of Engineering,
this is often called Continuing Engineering Education.

The fundamental goal of Continuing Education is to keep the
sum of competencies of an individual appropriate in time and
always relevant to the demands of scientific-technical evolution.

Two essential forms of Continuing Education are characterised
by differing trends: self-teaching, which is continuous, and institution-based, which is occasional. The latter is characterised by
generally shorter training schedules, which gives rise to the need
for a stringent schedule of Continuing Education courses [2].

4.4   Teaching

Flexibility for a teacher, whatever the course's level, is extraor-
dinarily important. A good teacher should constantly monitor his
students: there are many signs, many language manifestations,

especially non-verbal language, which can provide precious information for the teacher, who must be able to rapidly perceive them and change the teaching style, for example, by changing the speed of presentation or the tone of voice or the choice of examples. Subsequently, inviting students to give feedback can also provide important information.

It is also very important for the teacher or, generally, for a speaker, based on the characteristics of the audience, whether known or estimated, to artfully and appropriately adapt the style of presentation.

It goes without saying that also, and especially, for a teacher, staying up to date and, therefore, Continuing Education is of the utmost importance.

Three examples taken from the authors' experience can be significant:

(a) An expert biologist invited to speak at a conference on natural medicine at a Rotary Club, during an initial discussion at the pre-conference cocktail, perceived a certain opinion on the topic on the part of the audience, and quickly changed the approach she had prepared, making it more appropriate for the audience

(b) In a similar scenario, a famous surgeon did not heed warnings that the audience was not made up solely of doctors. Hence, he made his presentation without adapting the material to the audience, as if he were in a classroom with his students. He even accompanied the presentation with gory images directly taken in the operating room. Needless to say, the conference was a disaster. The surgeon was ready to hold that conference on the topic with that style, and would not have changed it even if his audience had been made up of children!

(c) An IT engineer, despite extensive experience and professionalism in a university setting, was invited to hold a seminar to students from an intermediate course. When requested to slightly simplify his presentation, he answered: *My seminar is made up of 800 slides and I can't change it.* Once again, it is pointless to say that the seminar failed in its intent, and it is better not to wonder what the audience retained of its contents!

These examples exhaustively show that flexibility is necessary for a teacher in any area and at any level.

References

1. http://jamesclear.com/eisenhower-box. Last visit 27 Aug 2020
2. S.D. Cova, E. Rovida, An Italian experience: the continuing engineering education program at Politecnico di Milano, in *5th World Conference on Continuing Engineering Education*, Helsinki, Finland (1992)

Ethics and Integrity

5

Ability to follow a given deontology.

This chapter presents large parts of E. Rovida and G. Susani's text "Deontology in Engineering Field. Proposal of an inter-regional code" published by the Order of Engineers of Milan, which we wish to thank for the kind permission granted.

5.1 Some Definitions

The word "ethics" derives from the Greek "ethos", costume, attitude, and the word "morals" can be considered as the Latin translation (in Latin "mos" means costume. Ethics (or morals) is, therefore, the science of costumes, of attitudes, of the way of behaving and acting.

According to [1] "morals" means the overall opinions, decisions and actions through which individuals express and recognize what is good and what is evil, whereas "ethics" means the critical reflection on morals.

The moral vision corresponds to the distinction between what the individual considers "good" and what he considers "evil". It must be observed that even "bad" people have a moral conscience: in fact, they follow what they consider to be "good" [2]. Let's consider, for example, two fully opposite persons, such as Mother Theresa and Adolf Hitler. The former had a moral vision leading to the identification of 'good' with helping the poor

© Springer Nature Switzerland AG 2022
E. Rovida and G. Zafferri, *The Importance of Soft Skills in Engineering and Engineering Education*,
https://doi.org/10.1007/978-3-030-77249-9_5

and the distressed, and she adjusted her life accordingly. Adolf Hitler, instead, had a moral vision that identified 'good' with the creation of a superior race, and his life was oriented accordingly.

The moral vision is, therefore, personal and needs an instrument of evaluation and correction: this instrument is ethics, which can be considered a rational and coherent system to determine good and evil [3]. The terms "morals" and "ethics", often considered as equivalent and, to a great extent, as synonyms, present this distinction: ethics is control and correction of morals. Source [3] reviews some aspects of ethics and points out some university courses on the subject, with particular reference to the methodologies used.

The conditions required by ethics, that is by the evaluation of the actions and of the relative behaviour, refer to:

(a) Reason: the subject must possess the necessary rationality to evaluate good and evil, as they are the subject of ethics; he must also possess the capacity to distinguish between good and evil in relation to ethics and not to the "moral conscience"
(b) Ideology, or the overall principles underlying ethics and pointing out what is good and what is evil; we are, therefore, concerned with natural ethics, Christian ethics, Marxist ethics and so on
(c) Free Will: it is obvious that after evaluating good and evil by using ethics as a tool, the individual must be free from any type of compulsion to act consequently.

Ethics can also be defined according to its own characteristics as:

(a) Teleological: it starts from the principle that an ethical action produces good, thus prevailing on evil; according to teleological ethics, the target of the action prevails on the subject's intention and on the means to reach it; it can be concisely expressed as "the target justifies the means"
(b) Deontological: when evaluating an action, the subject's intention must also be considered, since it is deemed to prevail on the target action.

Concerning the subject to be studied, we can still distinguish between:

(b.1) Theoretical ethics: it establishes the basis of the duty and deals with the laws of human actions and of conscience, considered as the ability to know the laws and to apply them to the circumstances.

(b.2) Practical ethics: it deals with duties and infringements.

We also speak of descriptive ethics (which describes the behaviour of human beings) and of prescriptive (which gives instructions about duties) and subjective ethics (which deals with the acting subject, irrespective of the actions and intentions), as well as of objective ethics (which considers the action relative to the common values and to the institutions).

The utilitarian approach to ethics must try to get the maximum advantage for the maximum number of people. It can be:

(a) Based on actions: the ethics of each action is evaluated according to a "for and against" criterion, that is in relation to the action itself

(b) Based on rules: the ethical rules are evaluated according to the benefits implied.

The ethical content and the consequent judgment of an action can vary according to certain circumstances. Some of them can be traced to the following:

(a) Ignorance, which can be referred either to the existence of the law or to the fact that a certain action is included in a particular law; it is always linked to the lack, sometimes faulty, of essential information

(b) Fear, which can affect both internal and external actions; fear can refer both to others' judgment and to the possibility of making mistakes, that is "to create a bad impression"

(c) Violence, which can influence external actions

(d) Education, which can distort the concept of "good" or "evil"

(e) Disease, which can press the subject to do actions of which he is not fully responsible, such as, for example, in the case of kleptomania.

The ethical level of a deed can vary in relation to the circumstances in which it happens. Some general rules are pointed out hereunder:

(a) An essentially good action can become better in terms of purpose; for instance, if someone works in the best way aiming at the community's advantage, this can be considered preferable to working in order to draw the boss's attention to himself

(b) A good or neutral action becomes negative if the target is negative; for instance, it is certainly positive to apply the correct rules of "design science", but it becomes obviously negative, if it is used to produce antipersonnel mines

(c) A negative action does not become good if the purpose is positive; for instance, working in a firm producing the abovementioned mines does not become a positive action if the engineer "has got a family" to support

(d) A good or neutral action with good and bad effects is allowed, if the good effects overcome the bad ones; for instance, not to pass a seriously unprepared student at the state exams has negative effects on the candidate and positive ones on the community, and the advantages for the community are greater than those for the student

(e) A good or bad action in terms of objective and aim can become better or worse according to the circumstances. For instance, working at one's best is certainly positive and can be even better if the aim is to give a positive service to the community. A further example is working superficially, which is undoubtedly negative and is even worse if this way of working is done to get a colleague blamed for it.

Deontology studies duties in relation to specific social situations. Deontology can also be defined as "work ethics", while "integrity" is related to the intents, with what you intend to do.

If concerned with a certain profession, deontology is always considered a synonym of "professional ethics" and, therefore, it makes up the whole of the behavioural rules relative to this profession. Deontology is consequently a concept that includes both (moral) ethics and professional ethics. The last can be considered as an individual's job where he has got a good level of qualification. The profession has some intrinsic fundamental characteristics [4]:

(a) It requires a remarkable training period
(b) The training has got an intrinsic remarkable intellectual component
(c) Knowledge and skills are essential to give the society a valid service
(d) It is characterized by a high level of decisional autonomy
(e) It requires an ethical rule.

So, this is the link between the profession and the ethics from which deontology comes.

In this book, the term "deontology" will, therefore, be used with this meaning; hence, as a synonym of "professional ethics".

Deontology is strictly connected with the subject's responsibility. In the field of engineering as well as in others, a good test of responsibility [4] can be the question: "Before acting in a certain situation, how should an engineer behave when there is nobody to control him?". Another criterion is to evaluate our behaviour by asking ourselves if such behaviour might be approved if another person were to behave in this manner towards us, following the old maxim: *do not do unto others what you do not want done unto you.*

5.2   General Criterion to Determine a Specific Deontology

A general criterion for determining a specific deontology relative to a specific profession can be the one deriving from the analysis of the "deontological space" (Fig. 5.1). It consists of three axes relative to

(a) Activities deriving from an analysis of the professional profile in question; for example in the field of engineering we can refer either to the academic courses or to the type of activity carried out

(b) Contexts of the activities and, in particular, persons or categories of persons the professional can come into contact with; generally speaking, they can be, besides themselves, colleagues, superiors, subordinates, customers, suppliers of products and suppliers of services

(c) Ethical principles, coming from the ideology to which they refer. So, if we refer to the Christian deontology, the ethical principles will be the evangelical ones; if we refer to natural deontology, they will be the natural ones and so on. The

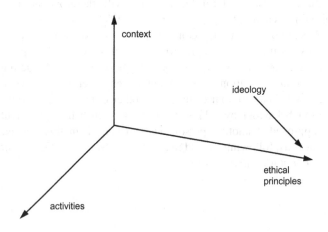

Fig. 5.1 Deontology space

points in space, the "coordinates" activities/contexts/princi-
ples, are the deontological problems, professional situations
where it is necessary to make ethical decisions.

Figure 5.2 represents a particular case of the application of
the general case. The ethical principle "to serve the truth" can
derive from the natural ideology. In the specific example, we can
consider design as human activity, while the context is the overall users of the product in the course of planning.

The point of "deontological space" characterized by the three
above-mentioned "coordinates" corresponds to the deontological
problem "to serve the truth when designing with regard to the
users of the product". An analysis of this problem leads to identify the duty to write the product instructions with the utmost
precision. Another example could be the one that refers to the
following coordinates:

(a) Ethical principle "Not to cause physical harm to the persons"
(b) Activity: "design"

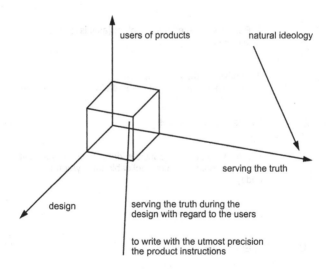

Fig. 5.2 Specific case to determine a deontology

(c) Context: "users of the product".

The deontological problem might be: *to apply the criteria of correct design with the utmost care to reduce the risk to a minimum when using the product. Moreover, to inform the users with the utmost precision and completeness about the residual risk and to identify the criterions of active and passive safety.*

5.3   Use of a Determined Deontology

Once a deontology has been defined, it can be used in the light of a concept of ethical cycle (Fig. 5.3).

For example, let's consider the case in Table 5.1, based on the professional experience of one of the authors.

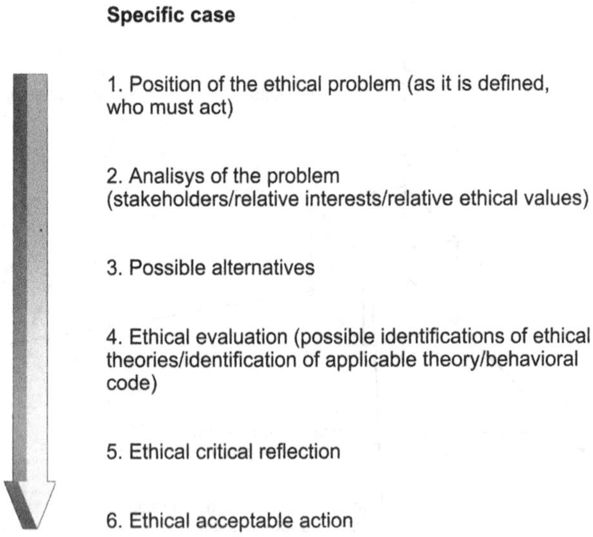

Specific case

1. Position of the ethical problem (as it is defined, who must act)

2. Analisys of the problem (stakeholders/relative interests/relative ethical values)

3. Possible alternatives

4. Ethical evaluation (possible identifications of ethical theories/identification of applicable theory/behavioral code)

5. Ethical critical reflection

6. Ethical acceptable action

Fig. 5.3 Schema of an "ethical cycle"

Table 5.1 Example of application of an ethical cycle

Phases	Situation
1. Position of the problem	A machine at the building site overturns and crushes a worker that immediately dies. The problem is to identify how the appointed Technical Advisor must act in a court of law
2. Analysis of the problem	The causes of crushing are recognized in the following points: (a) The machine is on a ground with a considerable transverse inclination (b) The container is full of thick concrete (c) The reaction couple caused by the drum starting up, summed to the overturning couple due to the transverse inclination of the ground. Moreover, but there is no evidence in this regard, the drum "starts up" suddenly The stakeholders are brought back to: (a) The machine's designer (b) The machine's driver
3. Possible alternatives	They can be traced to: (a) Design mistake: must the machine be designed and built in such a way as to stand up to the circumstances listed under 2? (b) Driving error: must the driver avoid the combination of circumstances listed under 2?
4. Ethical evaluations	(a) The designer's responsibility: he must foresee possible driving errors, but is contemporaneity really an "unpredictable" combination? (b) The driver's responsibility: can it be foreseen that, during site work, the driver can predict all those combined actions?
5. Ethical reflections	It seems that there are no clear responsibilities either for the designer or for the driver
6. Ethically unacceptable action	Since, in all honesty, both designer and driver do not seem to be at fault, the incident can be considered as caused by mere fatality

5.4   Some Examples of Ethics

5.4.1   Introduction

This chapter presents and briefly comments some "ethics" of different origins, since they can represent a useful introduction to the core point of this book, which is the engineer's deontology.

5.4.2   The Base of Roman Ethics

The base of Roman ethics, on which many legislations are founded even today, is expressed in the maxim of Enea Domizio Ulpiano (Tiro ca 170–Rome 228), one of the greatest Roman jurists, from whom Emperor Justinian the Great also largely drew for the Digesto. The maxim is the following:

> Iustitia est constans et perpetua volutas ius suum cuique tribuendi. Iuris praecepta sunt haec: honeste vivere, alterum non laedere, suum cuique tribuere,

which means:

> Justice consists of the constant and perpetual will to attribute his own right to everyone. The rules of the right are: to live honestly, not to harm others, to give everyone his own.

When reflecting on this maxim, which is so succinct, we cannot fail to admire the synthesis, typical of ancient Rome, and how it is the foundation of any form of justice. It is pointless saying that if everybody were to observe these rules, the world would be very different!

5.4.3   The Contribution of Rotary International to Ethics

Rotary International is an association that provides services. Established in 1905, and present in more than 160 countries with 1,200,000 members, it represents all the professions of civil society [5].

One of the members' targets is to reach the highest standards of professional ethics, for which some criterions of evaluation are hereafter pointed out:

The proof of the four questions is the fundamental point of Rotary ethics. It consists in asking oneself the following questions before any action. The answer could obviously be in the affirmative:

a. Is it true?
b. Is it right?
c. Does it improve interpersonal relations?
d. Does it represent an advantage for all people involved?

The Four Question Test was proposed in 1932 by Herbert Taylor, President of Rotary International in 1954–55, and was officially adopted by Rotary International in 1943.

5.4.4 The Hippocratic Oath

Even if it is not an argument that concerns the engineer's profession, we deem that a consideration about the Hippocratic Oath can be interesting, since it summarizes the medical professional deontology, in the modern version, strictly derived from the original one [6].

"Conscious of the importance and solemnity of the action I do and of the commitment I take on, I swear:

- To practice medicine freely and with independence of judgment and behaviour, avoiding any undue conditioning
- To pursue the defence of life, the guardianship of human physical and psychological health, and relief from suffering of which all my professional actions will be inspired with responsibility and constant scientific cultural and social commitment

- To cure each patient with the same care and diligence, irrespective of any race, religion, nationality, social and ideological condition, and to promote the elimination of any form of discrimination in the sanitary field
- Not to act in such a way as to cause a person's death deliberately
- To avoid any diagnostic and therapeutic perseverance
- To promote the therapeutic alliance with the patient based on confidence and reciprocal information, respecting and sharing the principles of the medical art
- To be true to the ethical principles of human solidarity against which I will never use my knowledge, in the respect of life and of the person
- To place my knowledge at the disposal of the progress of medicine
- To entrust my professional reputation only to my competence and to my moral qualities
- To avoid, even outside professional practice, any action and behaviour that can damage the dignity of the profession
- To respect colleagues even in case of different opinions
- To respect and facilitate the doctor's right to free choice
- To give urgent assistance to anyone who needs it, and to be at the disposal of the competent authority in case of a public disaster
- To keep the professional secret and to defend the discretion of all that has been revealed to me, which I see or have seen, understood or guessed in the practice of my profession or related to my condition
- To give, in science and conscience, my work with diligence, ability and prudence and with equity, observing the deontological rules of the medical practice and the legal ones that are not in contrast with the aims of my profession".

It would be interesting to study if the Hippocratic Oath can also be adapted to the engineering profession.

5.4.5   Ethics in Daily Life

The Four Questions Test can be valid in daily life. Just think how different the world would be if everyone, before every word and every action, checked if what they are about to say and do was true, fair, if it contributed to the improvement of human relations, and if it represented an advantage for all the people involved.

Unfortunately, each of us has a vast experience, too often suffered, of false claims made in bad faith, of even elementary violations of justice, of slander aimed at discrediting a person or sowing discord, uncalled for insults, words and actions that are intended to harm some people.

5.4.6   Ethics in the Engineering Profession

5.4.6.1 General Principles

Engineering is a wide and articulated profession, with a large number of facets. A general discussion is very difficult, but it is the authors' opinion that the Four Questions Test can apply to each of them. The design engineer will be considered here, as an example, also based on the experience of the authors.

The professional profile of the designer can be traced to the following definition [7]:

Starting from the function and specifications assigned (if necessary, correctly formalized)

- Using methods (design methodology, TRIZ, LCA, DfX, ...) and means (hardware, software, materials, components, ...)
- He determines all the information necessary to make an industrial product (shapes, dimensions, materials, tolerances)
- In such a way that the product has acceptable behaviour in all phases of its lifecycle.

Table 5.2 The four questions test as a base for the ethical conduct of an industrial designer

Questions	Duties of the designer
1. Serve the truth	(a) Great focus on professional training (b) Great focus on updating (c) Match the challenges with your competencies in the work you take on (d) Each statement (both technical and not) must correspond to the truth
2. Serve justice	(a) Honest behaviour towards all human beings (professional operators and users of technical products) (b) A particularly strict behaviour when acting as a court-appointed technical advisor in a Court of law (c) A high level of labour productivity that can become an example for others
3. Improved interpersonal relationships	(a) Avoid, beyond any legal boundary, any situation where danger or inconvenience to all users of technical products can occur in any stage of the product's lifecycle: production, distribution, usage, disposal (b) Avoid designing products that encourage excessive consumerism in users (c) The financial gain must not be higher, as a goal, than the effort to improve the product, in view of enhancing relationships between people (d) Be an example of professionalism, motivation, passion
4. Advantage for every human being	(a) Carefully consider all the effects that the product being designed can have on users, in terms of health, safety, education and ethical improvement (b) The advantage for users of the product must be at least as important as the financial gain for the designer (c) Always keep in mind that the attitude of the subject always influences others, to a higher or lesser degree

Table 5.2 shows an example of how a deontology of designers can emerge from the four questions test.

5.4.6.2 An Experiment

Ethics begins to find some space, which should be increased, in educational programmes. In this regard, this application of the Four Questions Test to a didactic experiment may be interesting.

A workshop organized by one of the authors [8] for the members of Rotaract (the Rotary youth association) engaged thirty young partners dealing in the following fields:

(a) Insurance
(b) Communication
(c) Health
(d) Engineering
(e) Architecture
(f) Economy and commerce
(g) Banks
(h) Finance
(i) Company consultancy
(j) Scientific research
(k) Human resources
(l) Marketing
(m) Entrepreneurial activity
(n) Journalism.

The young participants were invited to specify each of the "four questions", based on their own professional experience. The results are reported in Table 5.3.

As a comment, it can be observed that the young people showed interest in the experiment and stated that the Four Questions Test offered them much food for thought which, it is hoped, they will not forget soon!

It is also interesting to note that the teaching of deontology tends to find space in the engineering teaching programmes and, in general, in the engineering world [9–19].

Table 5.3 Application of the rotary "Four Questions Test" to specific cases obtained from the different professions

It is true?	It is right?	Does it improve interpersonal relations?	Does it represent an advantage for all
To take care of one's skills and professional training	To act correctly in all situations	To help socialization	To put the team's success before one's own success
To communicate rightly and autonomously	To be loyal, honest and respectful towards all	To facilitate the colleagues' work	To realize something with a useful aim for the mankind
To communicate effectively	To respect the laws	To cooperate with others	To create well-being
Objective control of facts and responsibilities, chiefly regarding the information sources and always verifying if what is written is right	To manage equally the human resources	To encourage the educational and cultural growth	To get different ways of thinking to reach an integrated knowledge
	To reflect well on one's own actions	To recognize others' skills in the team work	To contribute to the growth of the human and professional skills of all people involved
	One's own actions most be of the utmost transparency	Activity to reach the advantages for the working group	To meet all needs

5.4.7   Ethics of the Engineering Teacher

The activity of the engineering teacher, like any other teacher, is attributable to the following three areas:

(a) Research, which is production of new knowledge
(b) Teaching, i.e., transmission of knowledge and verification of the level reached by the learner
(c) Organization, i.e., contribution to the function of structures.

5.4.7.1   Ethics of Research

Research is the production and enlargement of knowledge. It is important to distinguish between pure and applied research.

Pure research can be defined, with Galileo, as "reading in the book of Nature": such research can, therefore, be identified with recognition of an existing situation; moreover, it is not "good", and never "bad". Ethical aspects can be involved by considering the means of the research: if a research study, for instance, needed the death of human beings, it would not be ethically acceptable, also if it were connected with new knowledge.

Applied research presents a different situation: such research, based on the results of pure research, upgrades human power and, therefore, can have two ethical aspects, because their applications can be either positive or negative. This character, however, particularly concerns the use, instead of the results. It is, therefore, necessary to propose ethics for users too.

5.4.7.2   Ethics of Didactics

Didactics can be defined as the transmission of scientific-technical contents by using a system of signs: didactics is also the formalization of given contents. An example of ethics for teachers can also be derived from the Rotarian "Four Questions Test" (Table 5.4).

Table 5.4 Application of the rotary "Four Questions Test" to the ethics of didactics

Questions	Contents	Form
True	Correctness Updating	Results optimization
Right	Congruence with requirements (of other courses and of professions)	Correct and objective evaluation of results
Amelioration of relations among human beings	Difficulties congruent with the student's level Give an example of ethical living, not only of technique	Adequate competitiveness
Advantages for all other persons	Objects and contents that are congruent with the requirement	Results optimisation

Hence, proposing an approach to a "teacher's ethics", it is:

(a) Correctness and updating of contents: such a requirement encourages the teacher to keep reliable and updated sources as reference point

(b) Contents congruent with the requirements: the aims and the contents of didactics shall be congruent with the requirements of both professional and other related courses

(c) Difficulties congruent with the student's level: the difficulties of didactics shall be congruent with the student's level, by taking into account the level of didactics and of what is taught

(d) Example of ethical life, not only of technique: a teacher should be an example in the human and deontological field

(e) Optimization of results: it is necessary to connect the objectives with the available time by using the most adequate methods and means

(f) Adequate competitiveness: it is important that the students work in adequate milieu, with good integration between competition and cooperation.

Table 5.5 Application of the rotary "Four Questions Test" to the ethics of the organizational activity

Questions	Contents
True	Correspondence of every statement to the truth
Right	Willingness to contribute to the function of structures based on personal competence
Amelioration of the relations among human beings	Maintaining proper relations with colleagues one is working with, without evading a task by saying: "Somebody else will, anyhow, do it"
Advantages for all other persons	Willingness to help colleagues Avoiding non-constructive criticism expressed for the sole purpose of discrediting colleagues Making the most of time dedicated to work, avoiding pointless words

5.4.7.3 Organizational Ethics

Table 5.5 presents some ideas of ethics applied to the teacher's organizational activity.

References

1. I. van de Poel, L. Royakters, *Ethics, Technology and Engineering* (Wiley Blackwell, 2011)
2. J. Rowan, S. Zinaichi Jr., *Ethics for the Professions* (Wadsworthy, 2003)
3. V.A. Lofthouse, D. Lilley, Teaching ethic in design: a review of current practice, in *International Conference on Engineering Design ICED2009*, Stanford, 24–27 August 2009
4. C.E. Harris, M.S. Pritchard, M.J. Rabins, *Engineering Ethics* (Wadsworthy, 2009)
5. https://www.rotary.org/en. Last visit 10 Jan 2021
6. https://en.wikipedia.org/wiki/Hippocratic_Oath. Last visit 10 Jan 2021
7. G.F. Biggioggero, E. Rovida, *Il profilo professionale del progettista.* Progettare 2(1997)
8. A. Pappagallo, C. Porcari, E. Rovida, *Un'esperienza sulla prova delle quattro domande.* Rotary 2040 3(marzo 2011)

9. G.Y. Ybarra, Ethics in engineering education worldwide, in *28th International Symposium on IGIP* (1999)

10. V.I. Solntsev, The engineering ethics course design for lifelong engineering education, in *30th International Symposium on IGIP Klagenfurt* (2001)

11. A. Can Ozcan, Ethics in industrial product design (Good, GOODS and Gods), in *International Design Conference DESIGN 2002*, Dubrovnik, 14–17 May 2002

12. G.F. Biggioggero, E. Rovida, Professional deontology of teachers in technical fields, in *33th International Symposium on IGIP in Cooperation with IEEE/ASEE/SEFI*, Fribourg, 27–30 Sept 2004

13. N.S. Nandagopal, Tools for teaching engineering ethics, in *33th International Symposium on IGIP in Cooperation with IEEE/ASEE/SEFI*, Fribourg, 27–30 Sept 2004

14. J.P. Domschke, Kompetenz als Ethische Forderung im Ingenierberuf, in 34th *International Symposium on IGIP*, Istambul, 12–15 Sept 2005

15. V.I. Solntsev, Engineering ethics contributes to engineering padagogics, in *29th International Symposium on IGIP*, Biel (2000)

16. I. Lebowitz, *Die Rolle der Ethik in der Ingenieurausbildung und Ingenieurbeuf* (IGIP 1989)

17. M.W-Martin, R. Schinziger, *Ethics in engineering* (McGraw-Hill, 2005)

18. K.K. Humpreys, *What every engineer should know about ethics* (Department of Electrical and Computer Engineering University of Cincinnati, Cincinnati, Ohio, 1999)

19. B. Allenby, Engineering and ethics for an Antropological planet, in *Workshop on Engineering Technologies and Ethical Issues National Academy of Engineering*, Washington (2003)

Relation with Others

6

After the above mentioned chapters, devoted to Soft Skills with a more general character, now some Soft Skills are examined, related particularly to specific aspects of relation with others.

6.1 Interpersonal Relationship

This Soft Skills can be defined as the ability to develop extraordinary relationships with others. Nobody works alone. Everyone is almost continuously in contact with a great number of people of equal, inferior or superior cultural, social or professional status. We could say this happens in all contexts, in both daily and professional life, whatever the occupation. Hence, we could say that everyone lives and operates in a group of people, be it large or small. So personal relationships play a crucial role, generally with the following objectives:

(a) achieve group results; they must be greater than the desire for personal gratification or the tendency to overcome others, which some group members might experience;

(b) mutual respect among the people who make up the group. This does not mean always agreeing and saying that things are going well, when they are not. When required to achieve the objective (a) it is necessary to be firm, while still respecting the individual.

© Springer Nature Switzerland AG 2022
E. Rovida and G. Zafferri, *The Importance of Soft Skills in Engineering and Engineering Education*,
https://doi.org/10.1007/978-3-030-77249-9_6

We can thus create a network of relationships where positivity is conveyed, establishing a harmonious atmosphere among group members, which allows to maximise the results achieved. This would require people who make up the group to view every situation in a positive light. Unfortunately, it is more common for people to see things negatively, to see problems in every opportunity. We need a shift in perspective, in other words, a paradigm shift: we need to see an opportunity in every problem.

How often, in the daily life, do we see groups of people who cannot reach a decision because some members of the group will not renounce their point of view (which might be clearly questionable!) to avoid giving other members the satisfaction of accepting their proposal. These are some examples of how interpersonal relations are carried out badly, or better, not at all. In many cases, this can also be the reason why families break up due to senseless rivalries, or because they cannot take ownership of their own mistakes.

In all of these cases, a consideration on the importance of correct interpersonal relationships can be very constructive for everyone involved, obviously as long as there is no malice involved.

If interpersonal relationships are important in daily life, they are even more so in the professional context, especially in the engineering field. It is precisely because of the sensitivity and the effort that engineering requires, on top of the fact that an engineer almost always works in a team, that perfect "teamwork" is required: harmony between the members of the group and, therefore, a great care for interpersonal relationships, are *conditio sine qua non (Indispensable prerequisite)*. And, in this case, a crucial role is played by the team leader.

Only too often we find an excessively authoritarian boss who creates an environment of fear, which is certainly not beneficial, both in terms of psychological wellbeing (and often even of physical wellness), but especially in terms of the team's results. This kind of leader (unwillingly, but often, intentionally) ignores the fact that the true job of a leader is (or should be) to identify each person's talents and the tasks that should be assigned, in order to optimize the overall group results.

In this case, and especially in this case, it is not necessary to be "Yes men" (which is actually negative and counterproductive). Always saying yes, perhaps to avoid annoyance or responsibilities, through laziness or for fear of other people's opinion, is a terrible way to relate to other people.

Good interpersonal relationships are very important for teachers in any field, but especially in engineering. A teacher must be able to appropriately relate to students, for example by creating a feeling of healthy competitiveness and creating interest in the subject they teach.

Instead, how many times do we see teachers who, far from stimulating students, instigate "hatred" for the subject they should be teaching students to love.

It is also important for a teacher, when assessing students, to know how to "exploit" their mistakes, through constructive criticism, which is often a great tool to overcome mistakes.

6.2 Positivity

Positivity can be defined as the ability to be optimistic and to convey positivity and positive feelings to the people around us. We all have some experience of people who convey energy, who smile, who take things the right way. After meeting them, we feel re-energized and satisfied. A positive person sees an opportunity in every problem although, it should be noted, such people are not the kind of shallow person who thinks and says something along the lines of "no need to worry, everything will be all right". Such people consider positivity merely an alibi for neither thinking nor taking action. Optimism, positivity, are not do-goodism at all costs. Indeed, an optimist also knows when, if necessary, to be harsh and yet respectful. The main feature of an optimist is the ability to see the positive side of every situation.

Positive people are generally quite proactive, in other words, in control. They can dominate their actions and, when faced with an unexpected event, they will first think and then choose the appropriate course of action.

On the other hand, there are people who are the exact opposite of this, who even at a glance can drain you of your energy. Criticising everything and everybody, they see a problem in everything. After meeting these people, you feel tired, drained of energy and empty.

Negative people are, broadly speaking, reactive. In other words, they tend to react automatically, without thinking and, therefore, they often act in an instinctive and erroneous manner.

There are even some people (and one of the authors knows something about this!) who, during a personal encounter, pretend to be friendly and flaunt politeness, while around others they do not waste a chance to make you look bad, sarcastically and mockingly, perhaps with a hidden (or not so concealed) desire to have you react in a way that will put you in the wrong.

PMA (Positive Mental Attitude) was first put forward by [1] and then revived by [2], who underline the importance of thinking positively and who see it as the beginning of success.

It is particularly significant how at the Yale and Berkley universities, there are very well attended courses on "positive thinking". This confirms the "hunger" for positivity there is in the world!

The general name of these courses is "positive psychology", defined as personal development and fulfilment, through the study of behaviours that foster happiness and the paths that lead to it.

Martin Seligman, Professor at the Pennsylvania University, says, "In the past we made an effort to make life less hard, while positive psychology aims to emphasize the most favourable aspects of ourselves, such as, for example, spirituality, positivity, skills, initiative and working on these aspects."

It is worthwhile to observe how we are all, to various degrees, the object of negative thoughts associated with people, facts or experiences. A common mistake is to attempt to drive them away. The more we chase them, the more they haunt us. The best thing to do is ignore them and leave them "behind us". One of the authors was told that negative thoughts are like thieves who try to break into a house. We need not go out and shoot them;

that would be counterproductive. It would suffice to not open the door!

We all have met positive, smiling people who do not make a mountain out of a molehill, people who know how to downplay problems, and to take the issues and inconveniences of life in their stride.

It is important to observe that a person's happiness (or, at least, peace of mind) depends on who that person is, not on what they have (a rationale of being, not of having). Happiness and unhappiness are within us. It does not matter what life throws at us, but how we "present ourselves" to life.

On the other hand, we all have a good idea, having met them, of sulky people who criticize everything and everyone, and see the negative in every circumstance.

A significant saying, that we should ponder, was often repeated to one of the authors by an aunt, "Those who lose money lose something, those who lose health, lose a lot, those who lose courage, lose everything".

Indeed, the way we "present" ourselves to life depends on courage.

In any work environment, a positive person simplifies issues, brings solutions closer and makes the environment better and more breathable. When faced with a technical problem, a "positive" engineer will be more inclined to see the strategies that can lead to a solution, and they will immediately think "I can do this!".

The opposite happens with a negative person, who only sees the dark side of events. They multiply and exaggerate problems, and make life impossible for both colleagues and subordinates. A "negative" engineer, faced with a technical problem, will only see the difficulties and immediately think, "I'll never solve it!"

How does a positive teacher act? They convey their passion for what they teach students, spreading positivity and hope, and showing students the best way to learn, to remember and to apply what they have learnt. They encourage students, they do not rebuke them for every mistake, and they recognize successes, saying, "Ok, you have done a good job and made it so far, now you can make a little effort and take a step forward". It must be noted, however, that a positive teacher is not one who promotes everyone (with less effort on their part), telling them that everything is going well when it is not. A positive teacher knows how, when necessary, to be strict, while always respecting the personality of students, recognizing the effort made by the student, and encouraging them to progress.

A negative teacher will do the contrary. They do everything possible to make the subject distasteful, assigning heavy and non-productive tasks, leaving out the interesting and useful part of what they convey (or should be conveying...). A negative teacher is a lead cloak. They divest students of motivation. They do not encourage them ("you have understood nothing", "you need to start over again", "you'll never pass the test") but even discourage their initiative.

It is quite evident that a negative teacher is destructive in any school, of any type and level.

6.3   Professionalism

Professionalism can be defined as follows:

(a) Ability to look and behave in a way that is congruent with a specific work environment
(b) Ability to update professional knowledge, in relation to the upgrading of the market's/environment's requirements.

Professionalism is, therefore, characterized by two aspects: the ability to acquire the necessary skills and the ability to maintain them adequate for the needs over time, in other words, training and updating.

In addition to the two above-mentioned aspects, the 10 Characteristics of Professionalism are also particularly interesting from the perspective of professionalism [3].

1. Dress for success
2. Confident not cocky
3. Do what you say you will do

4. Be an expert in your field
5. Behave morally and ethically
6. Maintain your poise
7. Have good phone etiquette
8. Strike the right tone
9. Be structured and organized
10. Own up mistakes.

A professional person worries, before taking on a task, about possessing the professionalism required. These skills are also expressed in many aspects of daily life, and here the discussion is closely connected to what was said about ethics.

A professional person is also balanced (here we see the chapter "Awareness"): he is confident without being arrogant, and precise without being fussy; he is punctual, without wanting to do everything with haste.

A professional person is not one who never makes mistakes (the best one makes the least mistakes!), but he is one who knows how to recognize his mistakes and, as far as possible, tries to remedy them.

The engineer's professionalism is obviously very important. In this regard, see many previous chapters. Here we mention, above all, the need to remain updated, which is an essential part of the engineer's professionalism. A very important component of professionalism is the ability to remain updated. A professional person knows how to promptly perceive changes in the market, in the environment and in the company, and is able to quickly adapt to them (also see the chapter "Flexibility").

Ability to update, means knowing how to quickly identify the "what", that is, the field where it is necessary to update and the "how", that is, the strategies and how to update. These can be:

(a) Self-study, such as, for example, books, magazines, conferences, software
(b) Institutionalized, such as, for example, courses, seminars and workshops organized by universities, academies, cultural and scientific institutions.

It is obvious that the professionalism of a teacher, especially in the field of engineering, is very important. Here, too, the components of professionalism are attributable both to the "what" to teach (the contents of the teaching, which must be continually reviewed and updated), and to the "how" (the application of the most effective teaching methodologies). In addition to this, there are many human aspects, which are referred to in other chapters of this book.

6.4   Responsibility

Responsibility can be defined as the awareness of personal tasks and duties, and the commitment to complete them satisfactorily. Responsibility is, therefore, strongly connected to ethics.

An analysis process leads to the highlight of the two constituent concepts: awareness and commitment. Awareness means knowing one's own abilities and also one's limits and also knowing that our actions, our words, and also our omissions have on the people around us.

Commitment means the ability to recognize one's mistakes and to bear the consequences. Each of us makes mistakes: the best and the one who makes the least mistakes! The difference between the responsible and the not responsible person is that the first recognizes their mistakes and tries to remedy them. The person responsible, then, whatever his field of activity, must give a positive example to all those close to him.

It is also closely related to the ability to foresee consequences of one's behaviour and to act accordingly. Responsibility, then, presupposes the freedom to decide about one's actions and, consequently, to be aware of one's behaviour. It is clear that no responsibility can be attributed to a subject who is forced to adopt a certain behaviour.

A subject must, therefore, be considered responsible for a certain situation when the following circumstances occur at the same time:

(a) the situation derives directly from the subject's behaviour;
(b) the subject could have foreseen, either with certainty or probability, the situation that occurred following their behaviour;
(c) the subject could have had the practical opportunity to act differently, thus preventing the situation from occurring.

Uncountable daily commitments enhance a person's sense of responsibility to various degrees. How many times do we witness, even in small things, someone who says, "I'll take care of it!" and then does nothing, either due to superficiality or because of other, more pressing commitments. Often the consequences of this attitude are very small, but this does not mean that, as a rule of thumb, this behaviour should be loathed. If for no other reason than that people who do not accept responsibility for small things will probably do the same for important matters too.

Every engineer, when carrying out their job, has many responsibilities, which can be considerable. Responsibility in the field of engineering is strictly linked to professionalism and ethics. Too often, in the professional field, in engineering, but not only, there are professionals who try not to take responsibility, but who try to blame colleagues for their mistakes

Responsibility, in this sense, has various levels:

(a) simply moral, when it exclusively violates moral duty and, therefore, the only penalty lies within the subject's conscience;
(b) legal, when it violates a legal obligation. In this case it can be:
  (b.1) civil, when the violation does not concern the penal code; for example, it can relate to a malfunctioning technical product that, due to an error in design, production, maintenance or use, can cause only a financial loss;
  (b.2) penal, when the violation concerns the penal code; for example, it might concern a malfunctioning technical product that causes physical harm to one or more people.

It is clear that we are faced with a combination of responsibility and deontology, and it is evident that an engineer must be well aware and sensitive to their responsibilities, not only legal, but also moral.

A teacher, in any context and at any level, has a great moral responsibility towards students. Suffice to think how they place their fundamental and unique resource, their time, in the hands of their teacher, who must endeavour to make the most of it. A responsible teacher, not only in engineering, but in every discipline, must always bear in mind that his task is not only to train professionals, but also to train people. Training good professionals is, especially, the transmission of Hard Skills, while training complete people requires a close and continuous integration between Hard Skills and Soft Skills.

In the field of engineering, but this also applies to other fields, the teacher has the institutional task of training engineers, but before engineers, they must train people. Hence, the teacher must put himself forward as a figure of reference, an example to students.

The fundamental task of an engineering teacher, but once again the topic can easily be generalized, is, as mentioned in other chapters, to generate passion for what they are conveying. It is crucial because passion, once triggered, can help to overcome any obstacle. Indispensable condition to awaken passion in students is for the teacher himself to be passionate about what he is teaching, only thus is passion transmitted.

Beyond generating passion, the teacher must also provide (or make an effort to provide) students with a method to face problems. Passion and method are two fundamental cornerstones of teaching and the basis of the "what" to teach, therefore the content of teaching itself. The "how" of teaching is also very important; in other words, the method used by the teacher to convey the "what". Communication, expressive skills and knowledge of didactic methodologies are crucial to the teacher.

All the above fall under the tasks of a responsible teacher.

6.5 Teamwork

Teamwork is an activity that is essentially based on mutual interaction between a group of individuals, and is aimed at achieving a previously established result.

Depending on the internal relationships between group members, we can distinguish between:

(a) Sociological group, characterised by individuals connected by their intention to pursue a common objective; some examples of sociological groups are a cultural association, a commercial company, or a political party;
(b) Psychological group, characterised by individuals connected to each other by a feeling of belonging to a super-individual entity. A sociological group can become a psychological one when the individuals who constitute it develop mutual feelings. The ties that are created between individuals belonging to a psychological group make it highly suitable for teamwork.

In terms of group training, we can distinguish between:

(a) Spontaneous training, consisting of the aggregation of multiple individuals who, almost by natural selection, converge to work together;
(b) Guided training, which entails an imposition from the team coordinator who, according to his own judgment, establishes which individuals are suited to be part of the team.

Teamwork methods can present considerable differences. The most important ones are described in Table 6.1.

The fundamental needs of teamwork include choosing the roles of the various team members, a very delicate job for the coordinator. It is, above all, important for everything that is done as teamwork to focus on fully achieving the result. This means that the coordinator must assign to each member the tasks he or she is best capable of completing. Moreover, it is crucial, though not always obvious, that the members respect each other, and thus we enter the specific field of certain Soft Skills.

Table 6.1 Examples of teamwork

Methods	Description
Cases	A real situation is simulated and presented to the group, which is then encouraged to: (a) Find a number of solutions (b) Study them, assessing them in relation to specified criteria (c) Summarize the chosen solutions
Carry out an event analysis	An "event", meaning a specific event, is presented Participants put forward a number of questions to the coordinator concerning the circumstances of the event A critical analysis of circumstances emerges from the discussion
Role playing	The coordinator presents a certain situation to the group The team members assign each other the roles and act out the situation
In basket	It consists of working on a certain number of documents, which are, generally, in the mailbox of the person, whose behaviour is under examination
Training—Group	It consists of free interactions between a certain number of individuals, without the guidance of a coordinator and without specific themes Often this serves as a preparation for subsequent teamwork
Business—Game	It involves the simulation of certain corporate functions, each of which is represented by a team member The resulting interaction aims at training decision-making skills to be used in companies and for a critical analysis of the related results
Brainstorming	The interaction of various people brought together without forewarning and invited to find solutions to a problem, which is presented unexpectedly

In everyday life, "teamwork" takes the shape of the family setup, group of friends, people who, for whatever reason, come together to make decisions. In such cases, rather than teamwork, we could refer to them simply as a group. Issues arising from these situations can very often be traced to the presence of someone who would impose their own will, whose main goal is to show off that he or she is clever, and to stand out for this, even making others look unskilled. Moreover, if the members are not strongly motivated, the group often does not achieve any result at all, and almost spontaneously dissolves.

Taylor and Fayol were considering, about the assumptions about People and Work, that the situation in organization where there was not just a distinction, between workers and management in terms of tasks, but also of education and social position.They believed in the necessity of detailed operational instructions and strong supervision as they considered that the uneducated work force required tasks broken down into simple components workers they were to do and be controlled and not to think and be empowered.

Managers and Workers did not talk sake to give orders and report back. Social intercourse between these groups was virtually unknown.

Douglas Mc Gregor , immediately after world war II, as a researcher at the MIT Sloan School of Management, considered that there were two positions that Managers could adopt when considering their work force: *Theory X and Theory Y.*

Theory X held that:
The average human being dislikes work and avoid working, if at all possible.

This dislike of work means that employees need to be controlled directed and even threatened, if necessary, if the organization is to fulfil its objectives.

People require direction but do not want responsibility.

Theory Y held that:
Work is a natural human function;
People relish responsibly;

The reward people seek are not only monetary;

The intellectual and creative potential of most employee is underutilized;

Toil is work that provides little reward for the person carrying it out.

Two of the most important writers on motivation, Abraham Maslow and Frederic Herzberg have stressed the importance of achievement and recognition. All of these are theory Y traits and are not particularly amenable to management by control. Thus the theory X approach is unlikely to assist in achieving the higher level of motivation.

The importance of motivation is very great. Indeed, humans are variable and thus those responsible for motivating people need to analyse the particular dominant motivational factors operating at any one time within the complex individual.

Following Mc Clelland , he suggested that human motivation was based on a balance between three needs: one basic psychological need such as food, water and shelter had been meet. The above mentioned three needs are as follows:

- The need for Power: for those seeking, supervisory and managerial position, power can be a considerable driver;
- The need for Affiliation: this can be allocated on a social implication of groups. While there are some who shun the company of others, most people instead wish to belong;
- The need for Achievement: one of the problems that this can cause is in relationship with others. Too high a need of achievement, at the expenses of the need for application, can lead to a very task-centered person, who is more concerned with achievement, than getting on with others. While this may motivate the individual, it can actually act as demotivation to others.

Very important is, also the demotivation calculus and psychological contracts. Each need has a strength and each reward can be quantified in how instrumental it is in reducing that need. The effort required will depend on both these factors.

This explains, in theory, why persistent failure produces a lowering of goals. If the need is high and considerable effort is expanded, but with little reward, it may be better to lower the need, if possible, and be sure of obtaining something.

Decision making, and behavior, with close links to motivation depends on weighing up of such situation based on need rewards.

Written contracts, because they are transparent, are easier to peruse, to see if there has been a breach by one or more of the parties. It is less easy to see whether first all the parties see the psychological contract in the same way and secondly if there has been a breach. Communication is all important and Tom Peters has coined the phrase "Management by wondering about and around". This means going around in the organization and speaking to both subordinates and colleagues in order to get an idea of their perception of the psychological contract which can be determined.

Coercive tends to be illegal in commercial dealing.

Calculative when both sides have something to gain although both may have to make compromises.

Cooperative is the ideal one in working, this happens when employees give maximum empowerment and are able to exercise high level of responsibility and self accountability.

It was an attempt to describe motivation with Japanese organization. This theory, called Theory Z, was characterized by offering life time employment (thus insuring a considerable degree of financial security and had subordinate/manager relationship based on mutual respect).

In late 1990 the downturn in Asian business growth hit Japan hard and organization fund they could no longer guarantee lifetime security.

Motivation is the key to understand team and leadership (better than management). Teams and not group (a gathering of individuals), while team is a group of single people. They can apparently have similar or primary reason but then the key issue to address is the relationship, if any between them.

The following criteria are important to get better relationship at work:

- Discover that your version of the "truth" may not be so true
- Do other people determine your emotions?
- Do you think your credibility is in question?
- How you tried to juggle to many roles at once?
- At the end of the day you feel: that nothing of real value has been accomplished
- Do your motives are aligned with your values and actions?
- Listen more, talk less, with the intent to understand better?
- Do you make it easy for people to tell you the truth?
- Do you think of humility as a strength or a weakness?

Teamwork is of the utmost importance in this field. The engineer rarely works alone but is almost always part of a team (in a company, in a firm, or in any other institution), in which harmony and the spirit of collaboration among team members is crucial to achieve results.

In short, the expected needs of team members can thus be summarized:

(a) Coordinator: they must be able to identify the features (in other words, the talents) of each team member, and assign tasks that are consistent with individual characteristics, so as to maximize the result of the task within the set time frame; it goes without saying that the coordinator should also make sure that the team's work environment is harmonious; this was already underscored when discussing other Soft Skills;

(b) Other members: they must strictly carry out the instructions of the coordinator, despite stating their opinions, which might be discordant. Moreover, it is important for them to contribute both to achieving the goal and to creating, within the team, a peaceful and mutually respectful environment. These considerations are also relevant to other Soft Skills.

Many aspects of teaching are team-based. Suffice to consider, especially in the engineering field, projects, papers, experiments and degree theses. Hence, please refer to the course to know about other Soft Skills where more examples of teaching with this type of set up are provided.

Figure 6.1 represent the importance related with the time of the relation with others; by upgrading the time of the relations, the increasing of the importance is corresponding to the progressive levels: contact, knowledge, rapport, esteem, trust.

Figure 6.2 is relative to the team typologies, in relation to the level of membership and the time of permanence.

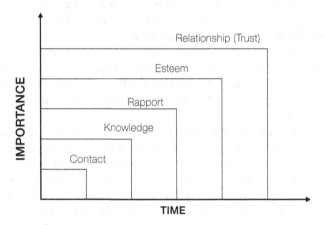

Fig. 6.1 From contact to relationship

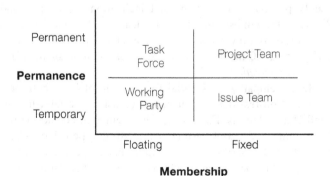

Fig. 6.2 Team Typologies

6.6   Empaty

The term derives from the Greek *en* (inside) and *pathos* (distress/ pain, considered as feeling/s). In practice, the empathic person recognises the feelings of others and steps into their shoes.

The empathic disposition is both intrinsic and acquired.

Empathy is difficult and complex because it concerns an inner impulse and the most vulnerable relationship we can have both with others and ourselves.

Citing the psychologist Carl Rogers, one of the first humanistic psychologists to explore empathic skills and their role in building human relations, *"empathy is the human ability to use appropriate communication tools, verbal, paraverbal and non-verbal (body language), to establish emotional contact with another person, identifying oneself with the other's subjective world with an approach centred on genuine and non-judgmental acceptance."*

Hence, an empathic person possesses the ability to change his perspective, non-judgmentally embracing that of another; he is, therefore, capable of recognising the other person's feelings. He does not limit himself to perceiving the actual situation only with his values, his beliefs and his paradigms, but recognises and respects the basic values of cultural differences and of different life experiences (social and cultural background). This is crucial in an increasingly interactive world. Empathy keenly perceives the feelings and inclinations of people, and this is of the utmost importance for team work, which requires the skill to both involve and share. Essential aspects for work in Engineering 4.0 include *team working, interaction, smart working, interconnection.*

Hence, empathy is essential for any role of life—both personal and professional—as it makes it easier to establish non-conflictual relations. Each of us daily comes in contact with a large number of people. An empathic attitude perceives others as neither adversaries nor competitors but as people in whom our empathic behaviour might leave a positive impression. This does not mean "being good" and accepting everything, or saying that "everything is all right" even when it is not the case. When necessary, the empathic person is also capable of being rigorous and even strict, but also of respecting others and of helping them understand their errors and correct them. Hence, empathy is an essential skill for interpersonal communications and a basic ingredient of human relations. It facilitates and enriches our relationship with others. Empathy is, therefore, considered a social skill.

Indeed, since empathy—an essential feature of emotional intelligence—develops several positive aspects, both emotional and relational, when it is correctly used, it can transform relations by removing blocks and making people more willing to listen.

Some feasible strategies to develop an empathic attitude:

- Practice "active" listening of the other person
- Learn to recognise and manage "feelings" (of self and others)
- Learn to establish emotional contact with others; this helps to understand them.

Observe the difference between empathy and sympathy: the former is an intentional attitude, while the latter, like antipathy, is unintentional. Hence, one can be empathic towards an unpleasant person.

There is also something called negative empathy, which must obviously be fought and overcome. It raises seemingly unremovable barriers between us and others. Some people "suffer" from negative empathy and see everybody negatively, judging them severely without distinction and condemning them without the chance of an appeal for their mistakes.

The empathic engineer, for instance, understands the feelings and the attitudes of colleagues, bosses and subordinates. He thus creates a pleasant work environment, office or workshop on a human scale. The outcome of the work thus improves considerably. Conversely, if the engineer is empathically negative, he behaves in quite the opposite way, criticising colleagues and subordinates in a non-constructive manner, ridiculing them and making fun of them. A person presenting negative empathy does not realise (or does not wish to realise) that, at times, even others can have their problems, either personal or family-related, and thus a negative approach to them only makes their life more difficult, finally impairing performance at work.

The importance of empathic relations between professor and student is evident as it facilitates learning and helps the student recognise and overcome errors, once again with considerable improvements to the outcome. Instead, if the professor behaves in a non-empathic manner (and one of the authors has experienced this!), he establishes an atmosphere of "terror". One can imagine how disturbing this can be for the educational relationship.

The important principles of *give* and *take*: why helping others can drive our success? For long time we focused that individual drive of success was the following: Hard Work, Passion, Talent and Luck. Today considering the dramatically reconfigured world success is increasingly depending on how we are able to interact with others and empathy is a fundamental issue to practice.

We can say that empathy is a SUPER POWER that makes the world a better place to live.

Figure 6.3 presents the power of empaty nicely figurative.

6.7   Engagement

Engagement is crucial for all activities and at all times. It presents two essential aspects. It can be active, when the subject involves other people by focusing on their talents (the brain's emotional hemisphere is very important in this case), on their passions and on their paradigms (and empathy is a valuable ally here). It must be said that engaging another person does not mean imposing decisions on others, at all costs. Engagement must always be based on respect for the person and his characterising talents.

What think feel?

What hear?

What see?

What say, do?

We can say that empaty is a Super Power that make the world a better place to live.

Just because you are right it doesn't mean I'm wrong, you just haven't seen life from my position.

Fig. 6.3 The empaty map

Then we find passive engagement, which is not less important. In this case, the subject allows himself to be involved in new initiatives, in paradigms as yet unexplored, in new ideas. Once again, we must underscore the fact that getting involved does not mean non-critical acceptance of the ideas of the "first passer-by" but, rather, processing every proposal both critically and ethically.

Engagement requires awareness of codes of conduct and of behaviour that are generally accepted in the various contexts and frameworks. Moreover, it requires understanding the cultural, social and economic dimensions of specific contexts, awareness of the subject's value and of his potential contribution in a given context.

Engaging also means expressing and comparing identity, points of view, needs and personal and context-related desires.

Actions and activities performed with a proper degree of engagement can actually "leave a lasting impression" on people.

Engagement can be either positive or negative. The former encourages virtuous behaviour, while the latter induces wrong behaviours. Hence, engagement must always be screened by ethical principles that, as a rule of thumb, lead to the exclusion of negative engagement.

It is also true that engagement, as mentioned, must leave an "impression", obviously a positive one, in the people involved. They must, anyhow, learn a lesson from engagement. Indeed, the etymology (from the Latin form) of the word "teach" means "impress" "in", that is convey something that transforms and leaves a deep inner impression.

Engagement is of the utmost importance for teamwork, whether it concerns a team, an office, a company or an association. It encourages people to share the principles, the objectives and the mission of the team they belong to. Healthy engagement strengthens the team, facilitating the achievement of objectives, without emphasising the effort required to achieve them. Indeed, suffice to think that a team's fate always moves through the people who form it. They are the most important component of every initiative; they are far more important than any raw material and any software!

Engagement is crucial in all moments of daily life. Knowing how to engage others and to be engaged even in small things makes all decisions easier and more productive. It is very easy to have experience of people that, even only in small things, are always of the opposite opinion, even if it is clearly wrong. Such people are resistant to be engaged, even in positive attitude. On the contrary, there are people that go to great lengths to engage others in negative behavior.

As mentioned, engagement (both active and passive) is particularly significant for teamwork. This leads to the essential role of engagement in the field of engineering that, as a rule of thumb, entails teamwork. An engineer must be able to engage his collaborators, transmitting enthusiasm to make the work on positive projects. Only in this way, he will be able to transmit passion to the students. And with passion, every obstacle is overcome.

A teacher who does not involve students, generating passion for the subject taught, is certainly not a good teacher. A teacher must also be capable of engaging with students, at least up to a certain point and with those who respond well to the teaching provided. To this aim, it is essential that a teacher has a passion for what he teaches himself:

6.8   Leadership

The Leader must be able to guide a group of people, motivating them to achieve common objectives. A leader must also be capable of delegating.

Becoming excellent leaders involves developing four key points:

1. Inspire trust
2. Clarify the objectives
3. Align systems
4. Release talent.

Trust has a positive effect on whatever the leader says. Low levels of trust have a high cost for the organisation, with the effect of a tax on relationships and a landmine for productivity and creativity, making it impossible, in practice, to achieve great results. A high standard of trust strengthens communication, facilitates teamwork and attracts followers.

Clearly define the goals, as they are the essential reasons for an organisation to have a vision, a mission, a set of values and strategies that should also be related to each corporate team's objectives (strategic connection).

Aligning systems means carrying out a process, a method and a set of procedures targeted at implementing priorities, allowing people to do their utmost, to be capable of working regardless of the leader's physical presence, and to look beyond the leader's presence in the organisation.

Releasing talent is one of the main duties of a leader who, besides being capable of providing support to achieve the objectives of individuals, thus enhancing their motivation, is capable of lighting the flame required to develop the talent that underpins a vocational calling.

A new era has to achieve great objectives by making the most of the highest levels of human genius and of motivation. Hence the need for leaders who follow new paradigms and possess innovative skills and tools.

What do excellent leaders do first? They see people as a single unit made up of body, mind and spirit, rather than consider them only as fixed costs. They must also be skilled in developing a successor, teaching him the "secrets of the job", and providing him their best guidance without fearing his success.

The leadership in the era of V.U.C.A. (e.g. Volatility, Uncertainty, Complexity, Ambiguity) world can be developed as follows:

Form the point of view of volatility:
Drivers: Change Nature; Change Dynamics; Change Rate and Speed
Effects: Risks; Instability; Flux
Demands: Vision: Take Actions; Probe Changes

Form the point of view of uncertainty:
Drivers: Unpredictability; Potential Surprises; Unknown Outcomers
Effects: Direction Paralysis Due to Data Overload
Demands: *Understanding*: Wider Understanding and Different Perspectives.

Form the point of view of complexity:
Drivers: Tasks Correlation; Multifaceted Effects; Influencers
Effects: Unproductive; Dualities
Demands: *Clarity*: Key Focus; Flexible; Creative.

Form the point of view of ambiguity:

Drivers: Ideal versus Actual, Misinterpretation

Effects: Induce Doubt and Dist Trust Lapses in Decision Making; Hurt Innovation

Demands: *Agility*: Decision Making; Innovation.

New rules needed for getting the right things done in difficult times. As a Team leader today you face an unprecedented challenge. This is not the time to reflect but to act, decide and energize the people around you with extreme urgency.

The hard true is that we have managed successfully during the prolonged period of good times may not be up to the challenges confronting them today. The Leader has to provide reassurance as well guidance to everyone through some aspects:

- *Recognizing reality*: there is simply no question that everything will be different two / three years from now.
- *Reallocating the true*: there is the need to interact with other more frequently and therefore the increased intensity requires a different time model matrix.
- *Protecting the Core*: it is necessary to identify the constituents and things that are the core of your activity and protect them from loss or damage during the crisis.
- *Being transparent*: the two major responsibilities, that must to be met, are the information flow and motivation. You must be prepared to listen to everyone, at any level. in order to discover what is in their minds, what they are worried about and what new ideas they have for doing something improved.

The most powerful of all competences for Leaders is the "inspiration". It is the quality most valued and is the factor most correlated with commitment of people and satisfaction. In addition, it will be important, that Leader action have to be bold or showing an ability to take risks confident and courage will also be necessary to explain how you reached your decision and why it could be the best decision to remain committed to a plan already in place.

The Paradigm of the Global Person, very important for a leader can be expressed as follows:

"SPIRIT - HEART - MIND - BODY"

Very important, also is the impact of inspiration: while inspiring others is a very critical competence. The important characteristics are communication skills and tenacity. Figure 6.4 represents how such characteristics can allow a possibility of positioning of the leader.

One important dimension of becoming a better leader is to clear about the outcome that have been shown to truly make a significant difference:

- Productivity
- Confidence
- Optimism and Hope
- Initiative
- Responsible behavior
- Enthusiasm
- Resiliency.

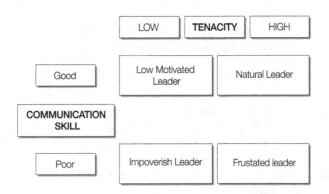

Fig. 6.4 Positioning of the leader in relation to communication skills and tenacity

Leaders inspire and motivate those around them determining a strong Contagious Emotions which fall roughly into two categories as follows:

1. New Behaviours and Outcomes
2. New Attitudes and Emotions.

The Leader is able to build this because all emotions are highly contagious from one individual to the next. The stronger the emotions that are expressed by the leader and the more willing and able the leader is to convey these emotions. The more change occurs within those being led. We can easily say that emotions are the DNA leader inspiration. It becomes essential therefore that inspirational leaders must be also excellent communicators and the most important key inspiring communication is that leaders are able to step into the listener shoes.

The final important step is the ability to develop people and fostering motivation.

References

1. N. Hill, *Think and Grow Rich* (The Ralston Society, 1937)
2. N. Hill, W.C. Stone, *Through a Positive Mental Attitude* (General Press, 1959)
3. https://smallbusiness.chron.com/10-characteristics-professionalism-708.html. Last visit 10 Jan 2021

Relation with Himself 7

In addition to others, the individual must also relate to himself. Here are some Soft Skills addressed to this type of relationship.

7.1 Learning to Learn

A key competence in both daily and professional settings is perseverance in continuing education. It is, therefore, essential to define a personal ongoing learning method. Motivation mainly stems from the need to constantly improve personal knowledge and skills. To this end, awareness of the importance of being competitive by always knowing the latest developments is essential especially in the professional setting. Being updated means aligning the learning goals with the evolving changes.

However, the intellectual pleasure of continuously learning new things should also be considered.

Indeed, the person who is capable of learning delights in acquiring new information. Moreover, he has a personal learning method, is aware of the challenges one might encounter when learning new things, and knows how to face such difficulties.

In order to learn to learn, one must be capable of self-evaluation, be aware of personal strengths and weaknesses, and identify personal interests and needs.

Why is it important to learn?

© Springer Nature Switzerland AG 2022
E. Rovida and G. Zafferri, *The Importance of Soft Skills in Engineering and Engineering Education*,
https://doi.org/10.1007/978-3-030-77249-9_7

Learning means acquiring ever new knowledge necessary to carry out the professional tasks required. Such knowledge can be described as constituted by the following components [1]:

(a) *Informative*, direct acquisition of notions (RUL)
(b) *Critical*, the ability to organise both analysis and synthesis of notions learnt, to express evaluations concerning certain opinions, and to organise the information acquired by key concepts and their classification according to both general and detailed links (EG)
(c) *Applicative*, the ability to apply the notions learnt and to critically process practical situations (EG°).

It is also advisable to perform a verification phase designed to compare the goals attained and those desired.

Table 7.1 is relative to the above mentioned components while the method is often called RULEG RUL + EG).

It is interesting to consider the Dublin Descriptors, as parts of the formation, particularly in the field of Engineering Education [2]. Such indicators are today widely used in engineering education and state that the student has to reach the following goals, during his/her education:

(a) Knowledge and understanding
(b) Applying knowledge and understanding
(c) Making judgements
(d) Communication skills
(e) Learning skills.

Table 7.1 The components of the RULEG configuration

No.	Component	Name	Abbreviation
1	Informative	Rule	RUL
2	Critical	Complete example	EG
3	Applicative	Incomplete example	EG°

It can be interesting the comparison between the above mentioned three components of the knowledge and the Dublin Descriptors:

(a) In the Dublin's Descriptors the knowledge (informative part, RUL) and the comprehension (critical EG) are considered together;
(b) "Makin judgements" can be considered a part of the comprehension ability;
(c) The "applying knowledge and understanding" ability of the Dublin's Descriptors can be regarded as equivalent to the applicative part (EG°);
(d) Communication and learning skills, with knowledge, comprehension and applicative aspects, could be considered as arguments to be directly taught, in specific courses, or, at least, in on purpose part of other courses.

As example, Table 7.2 represents the proposal of a specific course "Communication" for bachelor degree, with comparison of the Dublin's Descriptors and the RULEG method [3].

Table 7.2 Proposal of a specific course "Communication" for bachelor degree, with comparison of the Dublin's Descriptors and the RULEG method

Dublin's descriptors	Parts of RULEG configuration	Example of contents
Knowledge and comprehension ability	Knowledge (informative) RUL	Logical schema of communication and individuation of fundamental problems
	Comprehension (critical) EG	Ability to recognize the role of the communication in a given problem
Knowledge and comprehension ability applied	Application (applicative) EG°	Apply the above mentioned parts to a given simple practical situation

As further example, Table 7.3 represents the proposal of a specific course "Communication" for master degree, with comparison of the Dublin's Descriptors and the RULEG method.

Learning to learn means knowing the miscellaneous learning methods, styles and approaches.

Each person has a personal learning method. Feasible methods include:

(a) *Self-learning*, based, for instance, on reading books and magazines, on consulting websites, on participating in conferences and congresses
(b) *Institutionalised*, especially centred on attending courses organised by the various institutions the subject considers landmarks
(c) *Occupational (learning by doing)*, which allows to develop skills beyond their current limitations, in the workplace, a stimulating and also challenging context.

Additional learning methods:

(a) Rapid reading techniques, which allow to acquire more information in the same length of time

Table 7.3 Proposal of a specific course "Communication" for master degree, with comparison of the Dublin's Descriptors and the RULEG method

Dublin's descriptors	Parts of RULEG configuration	Example of contents
Knowledge and comprehension ability	Knowledge (informative) RUL	Structuring a general communication
	Comprehension (critical) EG	Ability to recognize the steps of the communication in specific cases
Knowledge and comprehension ability applied	Application (applicative) EG°	Realize a written communication and the correspondent oral presentation

(b) Writing, taking notes, drawing up conceptual maps. This makes it easier to identify gaps in the topic studied, and allows to better clarify ideas

(c) Identify objectives and key concepts, bonds and hierarchies between concepts, besides examples, the practical application

(d) Identify information sources by evaluating content reliability

(e) Create an archive, hardcopy and/or electronic, to check their constant validity in the framework of personal awareness.

Staying up to date, therefore, is very important for an engineer, as for all professionals, and in the field of Engineering, this is often called Continuing Engineering Education.

We can observe in Fig. 7.1 the performance over time of the training objectives (as mentioned earlier, often known as "terminal behaviour") required and effective throughout professional life.

The fundamental goal of Continuing Education is to keep the sum of competencies of an individual appropriate in time and always relevant to the demands of scientific-technical evolution.

Two essential forms of Continuing Education are characterised by differing trends: self-teaching, which is continuous, and institution-based, which is occasional. The latter is characterised by generally shorter training schedules, which gives rise to the need for a stringent schedule of Continuing Education courses [4].

In [5] some considerations about the Continuing Engineering Education, sometimes called Permanent Education.

7.2 Problem Solving

The problem solving can be defined as the ability to recognize and analyse problems aiming at defining targeted solutions.

In the current scene of great changes, organisations are often required to carry out their activities in situations that are often unforeseen, and which demand a proactive, rather than reactive, behavioural approach. In a "fluid" context, such as the current

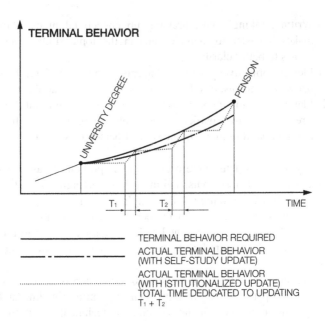

Fig. 7.1 Performance over time of required and effective training objectives throughout professional life

one, certain habitual activities often either prevent or do not facilitate proper problem-solving as they are outdated.

Identifying problems, defining and implementing effective solutions, even adopting innovative strategies, is the core process of problem-solving. It is important to define what to do (strategy), and to be capable of identifying the causes of the problem in order to eliminate and/or reduce negative effects by choosing possible alternatives.

At this point it is advisable to identify the necessary learning processes, and it is also essential to define information organisation models that can later be used in other circumstances.

Awareness of the context and of the process that is unfolding is important, besides knowledge of organisation and problem-solving methods.

After identifying the "problem", and if its high degree of complexity is ascertained, an analysis procedure should be adopted to divide it into a certain number of smaller problems.

It must be said that solutions can generally be identified in two steps:

(a) Known solutions, which are part of the general legacy of knowledge
(b) New solutions that can be triggered by identifying and applying innovative methods.

At this point, a choice of applicable solutions, identified based on the related strengths and weaknesses, can be formulated for each specific problem.

The important role of problem solving also in daily life is confirmed by the observation that often even in small everyday things, many people do not know how to act, precisely because they do not know how to identify the problem. This often starts with the inability to deal with the "what" to do, that is, the problem. And in these cases too often you see trying to deal with the "how". Thinking about how to deal with a problem without having clear what the problem is means making it very difficult to get to the solution.

It is also clear that the formalization of the problem correctly in the field of engineering is essential to arrive at the solution.

For a teacher, above all, but not only, of engineering it is then very important to identify the problem. There are two reasons for this. It is important, in explaining engineering problems, to make it clear to pupils which elements of a problem are important and which are less important. In addition, in the assessment of learning it is important that the teacher is able to identify the essential knowledge and those less essential for the evaluation of the student.

7.3   Digital Thinking

Digital devices have spread rapidly in a relatively short time, and any organisation that might present deficiencies in this regard would not be acceptable.

Awareness of the opportunity offered by digital devices is, therefore, essential, besides their informed and appropriate use. This means being proactive towards them and not experiencing them passively or, worse, hindering their implementation by saying, as at times we hear people remark, *But was it necessary?* or *Wasn't everything all right before?*

Hence the need to clearly understand the principles, know how digital devices function, and their rational and correct use. This generates the ability to perceive the characteristics of a context related to the use of digital devices, having a clear idea of their potential but also of the related problems and inconveniences.

This also triggers the need for high standard technical preparation designed to make the most of digital devices. However, this must not be distinguished from their ethical use. It is important to note that every technical innovation increases the power of humans and, therefore, presents ambivalent features. Subsequently, especially digital technologies require particular care to avoid any harmful use for both the person and others.

7.4   Planning

Planning means being capable of coordinating both personal work and the work of others, using time and resources to obtain effective results.

Planning signifies being familiar with the basics in order to manage activities and resources. In fact, the objective of obtaining effective results cannot disregard the proper and rationale use of both time and resources. Planning is linked to many other aspects of Soft Skills. For instance, defining a goal, as advised by the problem-solving process, is essential for planning. However, after establishing the goal, in this case we must take an additional step and "superimpose" a time axis to activities established by problem-solving, indicating not only the "what" but also the "how" to achieve the various objectives.

To date, only too often we see people, even those holding positions of responsibility, who lack a clear idea of the difference between what should be done immediately and what can be postponed. Subsequently, they waste time on pointless things (and words!!!).

Eisenhower's matrix is important to this end (see also the chapter "Flexibility"). It distinguishes urgent and non-urgent, important and unimportant items. It is also rumoured that the former President of the United States used this matrix to perform a huge amount of work. In fact, the matrix classifies all things to be done, based on their importance and urgency, according to the following categories:

(a) Important and urgent: these are the most critical items as they must be implemented rapidly and, despite their urgency, it is not easy to evaluate them as required (Do first)

(b) Important and non-urgent: they must be weighed, and there is time for this; hence, these items require and deserve the utmost attention (Do later)

(c) Unimportant and urgent: they can be delegated (Delegate)

(d) Unimportant and non-urgent: they can be postponed indefinitely (Delete).

Since nobody generally works alone, it is very important (and not always easy) to coordinate personal planning with that of colleagues and collaborators.

The importance of knowing how to plan every activity by establishing hierarchies of things to be done is never emphasised enough.

7.5   Result Orientation

When starting and implementing any activity, it is essential to always have a clear mental picture of the goal to be achieved. In this regard, determination and perseverance are of the utmost importance, besides effective performance of the tasks.

It can be helpful to define a Strategy (*what has to be done, which means doing the right thing*) and a Tactical Plan (*how to do it, which means doing things efficiently*) to identify possible issues and subsequent feasible solutions to achieve the intended result.

Some consideration is also necessary to estimate how often the right things are done (efficacy) and how often these right things are done well (efficiently).

Given that the following points are possible:

- Do The Right Things Well (DRTW), for example, to correctly complete the chosen task within the estimated time, and to precisely and promptly provide the information required
- Do The Wrong Things Well (DWTW), for example, correctly and rapidly perform a pointless task but present a well written report on a topic of scarce interest
- Do The Right Things Poorly (DRTP), for instance, perform the chosen work superficially and slowly; present an important report containing inaccurate information
- Do The Wrong Things Poorly (DWTP), for instance, suspend urgent work to perform a pointless activity without enthusiasm; present a poorly written report on a topic of scarce importance.

For further details on this topic, refer to the chapter "Flexibility"—in Eisenhower's matrix item—that underscores how to best manage the various activities. Clear ideas and lucidity are characteristics that should always underpin result-oriented decisions and behaviours.

Interesting can be some considerations about the logical connections between Result orientation and other Soft Skills, in particular Problem solving and Plannig. All three of these soft skills are related to the behavior in the face of a problem. In Fig. 7.2, is represented a general logical about the reciprocal interrelations between the above mentioned Soft Skills.

The starting point, in face of a general problem, can be considered the correct formalization of the problem (what to do, i.e. clarify the problem) by identification of the wanted goals.

Later, the Planning is corresponding to the "how to do", i.e. to the correct and complete individuation of the procedures necessary to reach the wanted goals.

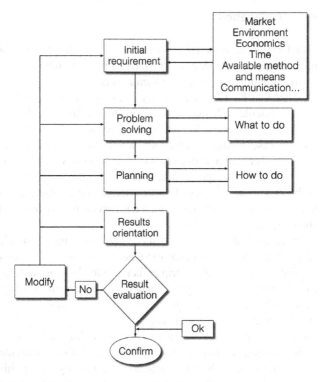

Fig. 7.2 Integration between problem solving, planning, result orientation

After the individuation of the solutions, the Result orientation is corresponding to the clarify of the results reached in the previous step.

The results of the problem should be evaluated in relation to some established criteria: the output of this step is the confirmation or the need for a change. Such operation can be relative to the Result orientation (new criteria to clarify the results), to the Planning (new procedures to apply, with the aim to reach the wanted goal), or to the Problem solving (new criteria to formalize the problem).

7.6   Awareness

In this field, awareness can be defined as the ability to criticise self, with the aim of adopting a balanced position between two opposite situations or feelings or mental states.

The ancient Romans used to say *in medio stat virtus* (virtue lies in the middle). This saying is just as true today, especially in the field of engineering. In particular, it can be said that a virtue or, in any case, balanced behaviour, when extreme, can become (or always becomes) a defect, sometimes with very negative results, especially in a professional setting.

Table 7.4 shows a series of virtues (in the first column) and of vices (in the second column), which correspond to a virtue taken to the extreme.

Confidence is a great virtue that can help overcome many difficulties, but when taken to the extreme, it can turn into a sense of infallibility, even a power trip, which can lead one to underestimate difficulties and to overestimate personal strength. Feelings of infallibility and power trips often underpin irrational behaviour, often with tragic consequences.

Fast decision-making, fast acting and the ability not to waste time in non-constructive hesitation, when taken to the extreme, turn into hasty and thoughtless actions. Often, by doing this, time is wasted, instead of being saved, as things done in this sloppy manner are either wrong or incomplete, and must often be carried out again.

Table 7.4 Balanced behaviours that can become negative

Balanced behaviour	Negative behaviour
Confidence	A sense of infallibility
Quickness	Overhastiness
Sharp wit	Abrasiveness
Alertness	Narrow focus
Dedication	Workaholism
Control	Inflexibility
Courage	Foolhardiness
Perseverance	Resistance to change
Charm	Manipulation
Ambition	Coercion
Power	Autocracy
Flexibility	Ambivalence

A sense of humour, inoffensive irony or a joke can often soften tension in a work environment, defusing critical situations. When it is excessive, the sense of humour becomes sarcasm and teasing, which often ridicules and embarrasses the target. While sharp and inoffensive irony improves the environment of a workplace, sarcasm, which is offensive to the target, makes the work environment untenable, worsening the results achieved by the team.

Vigilance, attention, constant and discreet tracking of what our co-workers are doing is a precious virtue, which increases team performance. However, if this behaviour becomes excessive, it turns into an obsessive breathing down collaborators' necks, creating tension, which is very negative to the state of mind of those working and, ultimately, to the quality and overall quantity of the work carried out.

Devotion to work is a commendable virtue, though it is, unfortunately, very rare today. Excessive devotion becomes a sort of work-addiction that, as all dependencies, can become very negative, even pathological.

The ability to lead is a valuable and indispensable virtue, especially for a manager who has (or should have!) the task of identifying each members' talents, and of assigning tasks that match these talents in order to maximise teamwork. The ability to issue orders, when exaggerated, becomes inflexibility that does not tolerate the slightest mistake, and is only capable of creating a climate of fear in the work environment. We can imagine how "useful" this would be for team results.

Courage is a virtue that allows us not to fear obstacles and challenges but, taken to excessive levels, it turns into madness that leads to useless risk-taking, with consequences that can be easily imagined.

Perseverance is positive behaviour, which helps us not to lose faith and not to be discouraged by difficulties and by inevitable failures. When extreme, it means being stubbornly attached to one's own ideas, not being willing to change attitude or behaviour, being "hostage" of one's habits.

Charm, within appropriate limits, can be positive and contribute to creating a pleasant work environment. If, however, a person "takes advantage" of their charm to manipulate others and impose their will, then charm becomes a questionable attitude.

Ambition helps us set higher goals and creates a spring that helps us improve. Too much ambition, however, becomes a compulsion, in other words, it runs the risk of becoming coercive or oppressive to those around us.

Power is essential in order to act decisively, but too much of it becomes authoritarianism, which is negative in terms of achieving harmony in relationships with others.

Finally, flexibility is a virtue that allows us to accept change, and to easily adapt to it. However, as stated in Chap. 4, an excess of flexibility becomes ambivalence, the condition in which we feel torn between two antithetic impulses, often both positive and negative, towards the same object. The negative effects of this situation, which greatly hinders any objective decision, are clear.

The comment to Table 7.4 underscores the importance of balance that should come from awareness and from objective self-criticism of one's behaviour.

A simple observation of the table immediately brings to mind how, in daily life, we often meet people who, despite being reputable and valuable in other ways, have no sense of measure. Through excess, they are capable of turning virtues into defects, which can make life harder for those around them.

Likewise, it should be pointed out that the environment of many workplaces could be different (and better), especially but not only, in the engineering field, if everyone were inspired by the table in question, and studied it with a sincerely self-critical spirit.

We can say that we have all, unfortunately, had some direct experience of teachers at all levels who, had clearly never seen… the table in this chapter. We can only imagine how this must have impaired the results of the teaching imparted.

7.7 Personal Branding

It is essential to know oneself and to be aware of the know-how possessed. Indeed, as also discussed in the various chapters of this book, awareness of personal skills is essential for success in the occupational world. Skills can be divided into two large categories. Hard ones, which refer to the profession's specific know-how, are generally discussed with sufficient detail in official training institutions, such as universities, particularly scientific and technical ones, which already provide thorough education and training.

Instead, concerning Soft Skills discussed in this book, they are scarcely included in institutional training. Indeed, their importance is growing, whether in the professional scene or in teaching or in daily life.

When writing this book, the authors took into account the importance of activating a tool to evaluate soft skills. In this regard, they perfected an evaluation system based on a certain number of questions, which were later managed by a dedicated algorithm.

The test is organised as described below. The goal for each of the soft skills discussed in this book is the behaviour triggered by reading the book.

Since a person's actions and behaviour closely depend on the knowledge acquired, the combination of these two factors is often called 'terminal behaviour'.

Terminal behaviour relative to each Soft Skill is then analysed, identifying, for each one, the basic knowledge that constitutes terminal behaviour. One question or, in some cases more than one question, concerns each of these basic knowledge factors.

The answers are assigned with a scale based on the Net Promoter Score (NPS), a management tool designed to especially evaluate the loyalty of a client towards a product or a service. It has been modified by the authors to meet specific needs.

The scale of the original NPS system has a range of 0–10, and designates a score of 0–6 to detractors, 7–8 to passive subjects, and 9–10 to sponsors.

In this specific case, the authors adopt the following division (Table 7.5):

(a) 0, 1, 2, 3, 4: low
(b) 5, 6: medium–low
(c) 7, 8: medium–high
(d) 9–10: high.

Following this, the authors perfected a specific algorithm capable of managing the answers and of providing an overall evaluation of the person taking the test:

Table 7.5 Evaluation and result qualification of the prepared test

Evaluation	Resulting qualification
0–4	Job holder
5–6	Actual best version of yourself
7–8	Unique person
9–10	Difference maker

The authors are at the disposal of whoever would like to receive, **through the Editor**, more information on the subject.

As example, here the questions relative to the Awareness are presented. Of course, the following questions are relative to the professional life, but it is easy to generalize the questions for the daily life, as well for the teaching activity.

Awareness

How to develop a good knowledge of one's.

1. In the professional field, I always try to assess the consequences of my actions, while making decisions quickly and trying to set higher goals, so as to continuously improve the work.

0	1	2	3	4	5	6	7	8	9	10

2. In the professional field, I am an ironic and charming person, in order to create a more pleasant atmosphere with colleagues and/or collaborators.

0	1	2	3	4	5	6	7	8	9	10

3. In the professional field, I am able to act decisively, but without excessive authoritarism.

0	1	2	3	4	5	6	7	8	9	10

4. In the professional field, I have dedication to my work, even if I take breaks, dedicating myself to my free time.

0	1	2	3	4	5	6	7	8	9	10

5. In the professional field, I am a person who can identify the talent of others and I know how to use it in order to put it in my work team to achieve a goal.

0	1	2	3	4	5	6	7	8	9	10

6. In the professional field, I am not afraid of obstacles and I do not allow myself to be discouraged by difficulties, while avoiding unnecessary risks.

0	1	2	3	4	5	6	7	8	9	10

7. In the professional field, I am a person who, while maintaining perseverance, willingly accepts change, easily adapting to every situation.

0	1	2	3	4	5	6	7	8	9	10

References

1. G. Gavini, *Manuel de formation aux techniques de l'enseignement programmé* (Edition Hommes et Techniques, 1965)
2. www.ecahe.eu/w/index.php/Dublin_Descriptors. Last visit 4 April 2021
3. F. Rosa, E. Rovida, R.Viganò, Proposal about the communication topics in engineering curricula, in *39th IGIP (International Society for Engineering Pedagogy) Symposium* (2010)
4. S.D. Cova, E. Rovida, An Italian experience: the continuing engineering education program at Politecnico di Milano, in *5th World Conference on Continuing Engineering Education* (Helsinky, Finland, 1992)
5. N. Torretta, U. Rossetti, Permanent Education: technical and scientific updating in engineering practice, in *Congress SEFI (European Society for Engineering Education)* (Aachen, 1976)

Imagination and Creativity

8

Ability to create something that did not exist previously (i.e. disrupting innovation).

8.1 Introduction

Imagination and Creativity are concepts that, despite a subtle difference in meaning, are characterised by considerable similarity.

By 'imagination' we mean the ability to conceptualize, identify and suggest ideas, solutions and new concepts capable of triggering a disruptive revolution in both markets and life. Creativity, on the other hand, leads to the transformation of ideas conceived through imagination, modifying the conceived and implemented innovations in order to improve them even significantly.

In terms of design theory, we can respectively speak of design for innovation (characterised by the creation of products that constitute a "leap forward", or a discontinuity in markets) and of design for change (characterised by improvements to already existing products).

In both cases, it means formalizing a certain function (innovative or already known, respectively) and identifying possible solutions that carry them out. The "best" of these must then be chosen, the one that carries out the intended function most consistently with the initial needs.

© Springer Nature Switzerland AG 2022
E. Rovida and G. Zafferri, *The Importance of Soft
Skills in Engineering and Engineering Education*,
https://doi.org/10.1007/978-3-030-77249-9_8

8.2   General Steps

The possible solutions can, generally, be identified by implementing the following steps.

The general objective is to identify the "best" solution to a given problem. Hence the need to recognise the possible solutions to the problem at hand. This can be carried out by following two steps in succession:

(a) Existing solutions, drawn from the individual's background knowledge, from the state of the art, from the indications required and/or received, from research carried out, for example, on the bibliography and/or the sitography. This step, as stated, especially concerns creativity, as it specifically consists in identifying improvements to existing solutions. These improvements can relate to both the solution per se (product improvement) and to the way of obtaining it (process improvement)

(b) Innovative solutions drawn, for example, from the subject's imagination (imagination is an important trait even in tech jobs!), from heritage (either taken as is or critically assessed), and from heuristic methods. Heuristics (from the Greek *heurisko,* literally "discover" or "find") is a part of epistemology and of the scientific research method that is involved in favouring access to new theoretical development, new empirical discoveries and new technologies [1].

Once all the solutions to the problem have been identified, the next step is to choose the one that carries out the required function in the most satisfactory manner.

Having hypothesised A, B, and C…, the various solutions, and 1, 2, 3… the different aspects revealed by functional analysis, the choice can be made with tables such as the one in Table 8.1.

The points in the table identify, with a numeric value or with an attribute, the value of the generic solution, compared to the generic aspect of the function.

If, as is often the case, all aspects do not have the same weight, this can be introduced by adding the following line to the matrix:

p1	p2	p3	....	...

8.3 Daily Life

The following is an example based on the personal experience of one of the authors. The task was to identify the person best suited to carry out a specific task. After identifying some people, of whom A, B, and C are three examples, the aspects that were deemed significant were the ones in the first column of Table 8.2, while the second presents the weight attributed to each aspect.

Referring to Table 8.2, Person A is the "best" fit: obviously, in this case, considering the limited number of people and aspects, the choice could have been made more directly. This is merely an example. It is clear that, if there were a higher number of people and of aspects, the use of the table would have allowed a more unbiased choice, with an innovative approach compared to a direct choice between few people. Thus, the shortlist to choose from can be wider.

8.4 Engineering

An innovative approach to a technical problem, instead of doing things the way they have always been done, can be to use specific methods, a few of which are exemplified below.

Table 8.1 General matrix for choice of solution

	Aspect 1	Aspect 2	Aspect 3	...
Solution A	VA1	VA2	VA3	
Solution B	VB1	VB2	VB3	
Solution C	VC1	VC2	VC3	

Table 8.2 Selection table for the most appropriate person to carry out a specific task

Aspects	Weights	Individual A	Individual B	Individual C	Weighted individual A	Weighted individual B	Weighted individual C
Age	1	3	2	3	3	2	3
Relationship with the person	1	3	2	3	3	2	3
Legal competencies	0.8	3	2	0	2.4	2.4	0
Living in Milan	0.5	3	2	1	1.5	1.5	0.5
Commitment compatibility	1	2	2	1	2	1	1

8.4.1   Historical Heritage

A critical analysis of the historical heritage can be a source
of interesting innovative ideas. Suffice to think that the
Conservatoire in Paris, opened in 1794, was suggested a century before by Descartes, with the aim of presenting machines in
chronological order, from the most ancient to the most modern at
the time, to technicians, in order to stimulate creativity and thus
encourage innovation. Descartes, therefore, had conceived the
embryonic idea of a database with the tools he had at the time.

To this end, a digitalised archive of implemented or even
only conceptualised ideas would be interesting. Such an archive
would be no minor instrument for innovation, based on a critical analysis of ideas from the past, actualizing them, for example
with modern materials and/or technologies.

As an example, [2] includes a proposal for an archive of constructive solutions relative to automobile suspensions. One of the
authors contributed to the project.

Another example is the wind car of Guido da Vigevano (about
1280–about 1349), born in Vigevano, Italy. Guido was Physician
and Engineer. As Physician, he was active at the French court.
As Engineer, he conceived a vehicle actuated by wind. The
remarkable aspect of such vehicle is the idea of the positioning of a windmill, with the revolving cap, on a two-axle wagon:
the rotation of the sails set in rotation, through a driveline, the
wheels, to allow the movement of the vehicle (Fig. 8.1) [3].

De modo faciendo aliud carrum quod ducetur cum vento/et sine
bestijs quod cum furore curret per campos planos pro confundendum
omne exercitum" "On the way of making another wagon which may
be propelled by wind/without draught-animals which runs on flat-
lands in order to confuse the entire army.

The idea of Guido to use a windmill (or, better, to use a modern term, a wind turbine) to power a wheeled vehicle survives in
the present, as sport practised in desert or beaches: of course, the
used technology is modern.

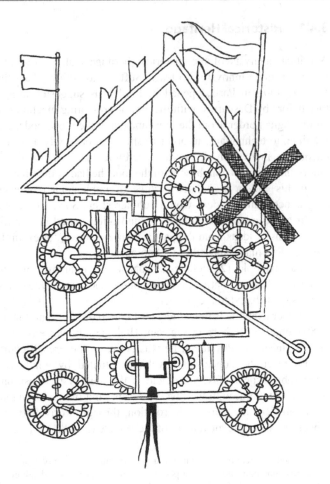

Fig. 8.1 The first drawing of a wind wagon, by Guido da Vigevano (Guido da Vigevano (ca. 1280–ca. 1350), Texaurus Regis Francie, 1375, Yale Center for British Art, Paul Mellon Collection, Mil mss (4°)

8.4.2 Traditional Heuristic Methods

Innovative solutions can emerge from the consideration, for example, of physical phenomena. We can consider designing a technical product by changing the physical phenomenon used to date to carry out the function.

Constructive solutions, which differ from those known, can also emerge from, for example, geometrical, kinematic and dynamic variations. Table 8.3 presents a few examples.

As example, Fig. 8.2 [4] is relative to some principles realizing the transformation of the continue rotating motion into alternative translation motion.

The analysis could be continued. For example, if the third principle of the figure is chosen, the component functions that can be highlighted, among others, are the following:

(a) Contact between moving and moved members;
(b) Return of the moved member.

The second one can be realized by means of weight (Fig. 8.3), by means of spring (Fig. 8.4), or by means of gas pressure (Fig. 8.5).

Another possible variation can be to change the configuration of surfaces around functional surfaces, for example by changing materials and/or technological processes.

Figure 8.6 shows an example of variation in the configuration of a bell-crank lever, which can be considered the materialization of three holes having a given diameter and a given centre-to-centre distance, and arranged according to a specific angle.

It can be seen that the surfaces of the holes can be "connected" by a certain quantity of material, shaped according to specific technological procedures. For example, in the figure, the technological processes used can be, from top to bottom, respectively:

(a) From an indefinite semi-finished product, for example "cutting out" a portion of sheet metal;
(b) By fusing or shaping plastic, depending on the chosen material (not all materials are suitable for both technological procedures mentioned);
(c) By connecting parts obtained from indefinite semi-finished products (in the case at hand, the piece is obtained from three sections of pipe and from two parts of sheet metal, welded together);

Table 8.3 Possibility for different variations

Geometry	Forms Combination of forms Structural link Number Position Orientation …
Kinematics	Direction of motion Reciprocal positioning of mobile and fixed points Position and number of fixed surfaces Position and number of mobile surfaces Type of trajectory …
Dynamics (for example, force generation)	Weight Cable or chain traction Bar thrust Spring thrust Archimedes' thrust Fluid-dynamic pressure Friction Electrostatic attraction Magnetic attraction …

(d) From an indefinite semi-finished product (a similar solution to 1), where the sheet metal portion has a different shape).

For somewhat complex technical products, especially if made up of multiple pieces, a possible criterium to identify innovative solutions could be the following steps:

(a) Identify the general function carried out by the overall product

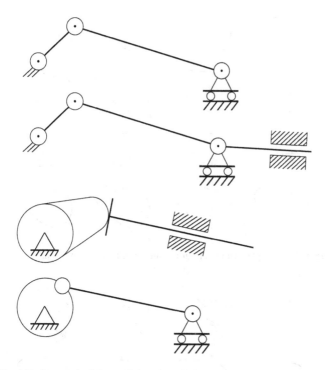

Fig. 8.2 Some principles realizing the transformation of the continue rotating motion into alternative translation motion

(b) Identify the component functions by analysing the general function

(c) Identify constructive solutions for the component functions

(d) Identify constructive solutions related to the general function, as a congruent synthesis of the solutions relative to the component functions.

For example, if we need to design a cylindrical hinge as per Fig. 8.7 [5]: a rigid horizontal axis is secured to a wall and a vertical arm is bound to it with a rotoidal joint. The movements this

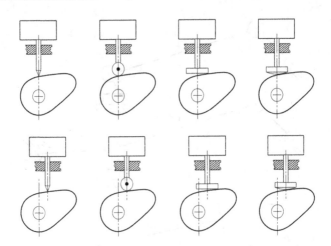

Fig. 8.3 Principles with the return of the moved member by weight

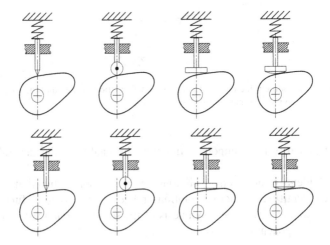

Fig. 8.4 Principles with the return of the moved member by spring

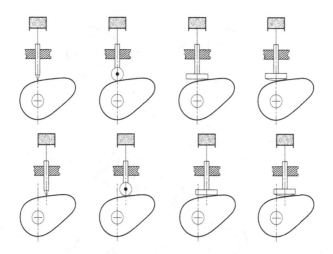

Fig. 8.5 Principles with the return of the moved member by gas pressure

arm can make are only rotations around the fixed axis, as axial sliding must be prevented.

The following component functions issue from an analysis of the general function:

(a) Lock the axis
(b) Allow the arm to rotate in relation to the axis
(c) Prevent axial shifts of the arm.

A rational and, therefore, innovative approach might be to find, for each component function, the construction principles/ solutions that can be used to carry it out, rather than doing what has always been done.

Table 8.4 provides an example.

The solution for the mechanism is a congruent synthesis of a single solution for each function (Fig. 8.8) [5].

The first overall solution to the above Fig. 8.8 is described below:

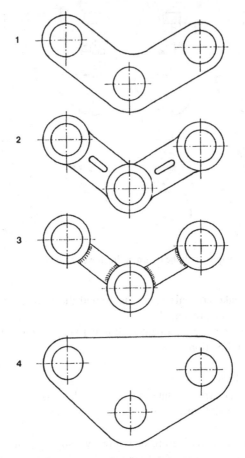

Fig. 8.6 Various configurations of a bell-crank lever. Courtesy of CittaStudi

(a) Axis lock: through threaded shaft with relief to allow securing on the surface with highest diameter; note that a cylindrical flat head pin is used to allow tightening
(b) Allow rotations: roller and cage assembly
(c) Prevent axial sliding of the arm: sliding to the right is prevented by the fixed wall, and to the left by the pin head.

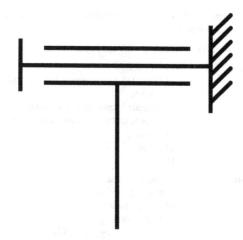

Fig. 8.7 A diagram of the mechanism to be designed. Courtesy of McGraw Hill Education Italy

The second comprehensive solution is given below:

(a) Axis lock: through a threaded shaft secured to incomplete threads; the stem head is cylindrical, with a hexagonal groove for tightening
(b) Allow rotation: through direct contact between the smooth part of the stem and the arm bore
(c) Prevent axial sliding: sliding to the right is prevented by the fixed wall, and to the left by the pin head.

Finally, the third solution is described below:

(a) Axis lock: by interference connection onto a stem with smaller diameter
(b) Allow rotations: two smooth flanged bearings mounted in opposite directions
(c) Prevent axial sliding: sliding to the right is prevented by the flange of the right bearing, which leans on the fixed wall;

Table 8.4 Examples of construction principles/solutions for each component function mentioned above

Component functions	Examples of possible solutions
Lock the axis	(a) Pressure (b) Threaded connection secured to incomplete threads (c) Threaded connection secured to a surface perpendicular to the axis and relief (d) Welding (e) Gluing (f) …
Allow rotations	(a) Direct contact (b) Rolling bearing (c) Journal bearing (d) Elastic bearing (e) Roller and cage assembly (f) …
Prevent axial sliding	(a) Cheese head of part (b) Hexagon head (c) Cylindrical flat head (d) Expansion ring (e) Diametral pin (f) …

sliding to the left is prevented by the flange of the left bearing, which leans on the pin head.

These are only three examples: there could be many more by combining the solutions exemplified in Table 8.4.

8.4.3 TRIZ

TRIZ [6–8] (Theory for inventive solution of problems) is a general method articulated as a set of tools with the aim of encoding the creative process in the technical field. TRIZ, which starts from the analysis of a large number of patents, is based on a few essential observations:

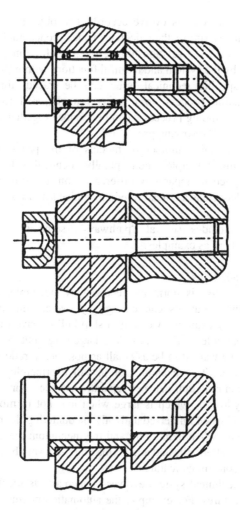

Fig. 8.8 Three examples of an overall solution for the mechanism, through a congruent synthesis of a single solution for each function. Courtesy of McGraw-Hill Education Italy

(a) Technical systems evolve according to objective laws, and tend to maximise their degree of ideal conceptual features, expressed as a relationship between useful functions provided by the system and harmful functions within the system.

(b) Any specific technical issue can be traced, through an abstraction process, to a general model, and the logical solution-finding processes can be grouped together in a finite number of "resolving principles".

(c) Given a finite number of models of the problem and of resolving principles, conceptually identical solutions can be applied to apparently different technical problems. Thus knowledge has a crucial role in inventive activities.

TRIZ includes several "pathways", some of which are described here as examples.

(a) Contradictions

Compromise is normally accepted as inevitable: the most effective solutions can emerge by overcoming contradictions. For example, we want a ship hull to have the biggest hull possible, which can hold a large cargo but, at the same time, we want it to be as small as possible to reduce motion resistance. Two examples of inventive principle related to this contradiction can be: separation in the course of time, the hydrofoil (which is large when it is not in motion, and "small" when immersed and only encountering the resistance of the air when in motion); and fragmentation: the catamaran (two smaller hulls and a large loading platform above them).

(b) Ideal conceptual features

Each technical system evolves to increase its ideal conceptual features. For example, the automatic opening and closing mechanism for the roof of a greenhouse in relation to temperature can be created with a hollow roof containing fluid, which manifests as a liquid below a certain temperature and as a gas above it.

(c) Resource management

It is important to consider all that surrounds the system, understanding even what can be considered harmful, such as, for example, production waste, heat to be dispersed and resonance.

(d) Space–time interface

The problem must be considered from all possible angles, for example by changing the perspective of a system, either observing it "from afar" or examining the smallest detail, monitoring how it evolves in time (examining the past and the future). For example, Table 8.5 shows how the "locomotion on snow" function can be examined [9].

The table is also remarkably inspiring for innovative solutions, precisely service as an alternative to the product.

8.4.4 Choices

Creativity can also arise from a criterion of choice of principles (or, respectively, of constructive solutions) identified according to the criteria exemplified in the previous sections.

The solutions, both in terms of component functions and in general, can be chosen by referring to the product's lifecycle (Fig. 8.9) [10] proposed by Asimov and commented in Chap. 9.

It means examining how these solutions "travel" the lifecycle, keeping in mind the project's requirements.

Referring to Fig. 8.9, the aspects arise from the product's lifecycle (i.e., production, distribution, disposal), while the weights correspond to the needs of the project.

The choice can be made by using tables such as 8.6, for example, which refers to the centring systems between two mechanical parts.

The example in Table 8.6 shows, as can be seen, assessments of the behaviour of constructive solutions for aspects expressed through attributes.

It is generally preferable to express the score in a numerical form: Table 8.7 shows one example related to constructive solutions for threaded fasteners.

The assessments in Table 8.7 are expressed numerically, hypothesizing 5 = maximum value, 1 = minimum value.

Table 8.5 Analysis of the "locomotion on snow" function from various perspectives

	Future	Present	Past
Supersystem—environment	Prevent harmful events (e.g. Reducing snow falls, providing alternative communication means)	Road, snow, communication infrastructures	Act on the environment (e.g. Removing snow, warming roads)
System—vehicle	Prevent accidents with the car (e.g. Taxi service, snowmobile sharing)	Car, motorbike	Act on the wheeled vehicle (e.g. Transporting the car on a truck/train, towing with a rope)
Subsystem—wheel and device	Prevent low-friction situations(e.g. Preliminary installations/replacement of worn winter tires)	Tire, tread, chains, spikes	Act on the tire (e.g. Renting winter tires/snow chains/spiders

Numerical assessments are particularly meaningful as they lend themselves to quantification and, in more complex cases, to the application of an expert system.

The problem of choices can be examined more deeply by applying the QFD (Quality Function Deployment). The starting point is the client/user's needs expressed as HLO (High Level Objectives). For example, considering the function "locomotion of a road vehicle on snow", the HLOs could be traced to the following [9] where the HLO are corresponding to the columns of the Table 8.8:

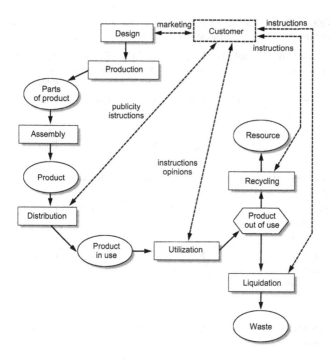

Fig. 8.9 Asimov's proposal of product lifecycle (see also Fig. 9.1)

(a) low price
(b) low installation-related costs
(c) low usage-related costs
(d) high safety and performances
(e) low environment-related costs.

The HLOs must then be "translated" into FR (functional requirements) deduced from the lifecycle. Table 8.8 presents an example.

Table 8.9 shows the weighted score of some constructive solutions for locomotion on snow, based on FR.

Table 8.6 An example of selection table for constructive solutions concerning the centring of two mechanical parts

Constructive solution	Simplicity	Relative rotation consent	Radial symmetry	Pressure potential	Adaptability to large dimensions
Cylindrical surface + plane	Good	Yes	Yes	Good	Poor
Conical surface	Moderate	Yes	Yes	Moderate	Poor
Two pairs of planes	Moderate	No	No	Moderate	Poor
Stud forced into the lower part	Poor	No	No	Poor	Good
Threaded pin fixed into the lower part	Poor	No	No	Poor	Good

Table 8.7 Assessment table for constructive solutions for threaded fasteners

Constructive solutions	Simplicity	Ease of assembly	Potential for repeated dismantling	Cost-effectiveness
Screw	3	3	1	3
Bolt	5	2	3	5
Trapped screw	2	5	5	2

8.5   Teaching

It is not easy to teach creativity. Indeed, it requires qualities that are mostly innate, such as imagination and power of observation. Nevertheless, some creativity exercises can be suggested. Three case studies will be examined herein, from the teaching experience of one of the authors [11].

Table 8.8 "Translation" of High Level Objectives (HLO) in Functional Requirements (FR)

FR	Price	Installation	Usage	Safety & performances	Environment
Final price in the stores	3				
Price/km	5				
Time to install		5			
Time to disassembly		3			
Cost to install		3			
Cost to disassembly		3			
Manual intervention		1			
Maintenance cost/km			3		
Space needed for storage			3		
Comfort			5		
Versatility/ flexibility of the device			5		
Maximum acceleration				3	
Maximum speed on snow				3	
Maximum speed without snow				1	
Lateral grip				5	
Braking performances				5	

(continued)

Table 8.8 (continued)

FR	Price	Installation	Usage	Safety & performances	Environment
Time needed to act				5	
Pollution produced					5
Damage to infrastructures					5
Recycling					3

Figure 8.10 shows a logical diagram related to the two courses held by the author: "History of mechanics" and "Design methodology"; together, they can inspire creativity. In the first case, students, after choosing a technical product, were asked to present, as project for the year, the reasoned historical evolution of such a choice. A critical analysis of historical evolution could have provided some innovative ideas, in some cases. Later, during the second course, students were asked to carry out research on current solutions for the same product: this research was carried out in steps, precisely, known solutions, by exploring catalogues and patent archives, and the search for innovative solutions, by conducting an investigation based on heuristic methods [12].

In the Machine Design course held by one of the authors, some exercise to upgrade the creativity of the students was utilized. Some examples are here proposed. In the Fig. 8.11 a connection between tank and lid with bolt is proposed. Such connection is presented in orthographic projections. The students, starting from such drawing, should represent another solution of connection (e.g. with screw or trapped screw) and different configuration of the lid, by considering another manufacturing process.

Table 8.9 Weighted assessment of some constructive solutions for locomotion on snow, based on FR

FR	Weight	Spiders	Snow tires	Autosocks	Snow claws	Put & go	Snow chains
Final price in the stores	3	1	1	3	3	5	1
Price/km	5	3	5	1	3	3	3
Time to install	5	3	3	3	3	3	1
Time to disassembly	3	3	3	3	3	3	3
Cost to install	3	3	1	5	5	5	5
Cost to disassembly	3	3	3	5	5	5	5
Manual intervention	1	3	5	1	1	1	1
Maintenance cost/ km	3						
Space needed for storage	3	3	1	5	5	5	3
Comfort	5	1	5	5	3	3	1
Versatility/flexibility of the device	5	1	5	1	3	3	1
Maximum acceleration	3	3	3	3	1	1	3

(continued)

Table 8.9 (continued)

FR	Weight	Spiders	Snow tires	Autosocks	Snow claws	Put & go	Snow chains
Maximum speed on snow	3	3	3	1	1	1	1
Maximum speed without snow	1	1	5	1	3	3	1
Lateral grip	5	1	3	1	3	1	5
Braking performances	5	5	5	3	3	1	5
Time needed to act	5						
Pollution produced	5						
Damage to infrastructures	5	1	5	5	5	5	1
Recycling	3	3	1	3	3	3	3
Total		145	203	181	197	183	159

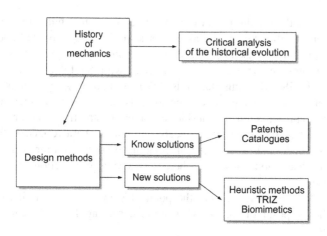

Fig. 8.10 Logical diagram to teach innovation by researching innovative solution

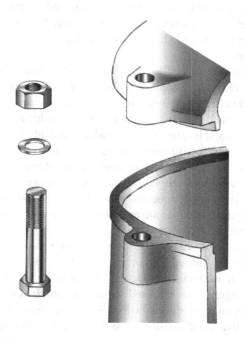

Fig. 8.11 Connection tank-lid to be studied with different constructive solutions. Courtesy of Cittastudi

Another valid exercise is relative to a shaft-hub connection. To the student the axonometric projections, mounted and exploded, is presented. The students are invited to represent such assembly in orthographic connection, with careful application of the ISO drawing standards this is a valid drawing exercise. But, after such application, the students are invited and to repeat the representation by considering other constructive solution of the shaft-hub connection. Each solution is characterized with the functional aspects and, in relation to specific requirements, the discussion above the choices is a valid completion of the exercise.

Very interesting, from the point of view of the creativity, is the proposal of innovative solutions, starting from the know ones.

Another valid exercise is shown in Fig. 8.12. To the students are presented the schema of the figure, at the top. Such schema is relative to a shaft on two supports. Right-hand support prevents horizontal shaft movements, while left-hand support allows shaft elongations and shortenings, due, for example, to thermal expansions.

Students are invited to draw an assembly design, choosing suitable constructive solutions. The figure below represents the drawing made by a student, still with numerous errors. The figure represents the shaft supported by two cylindrical roller bearings. Also in this exercise, the students should consider (and represent) new constructive solutions for the bearings, axial locking for the bearings and, of course, of the shaft.

The drawings of the various students are examined and discussed. Errors are commented and corrected, while the choices of constructive solutions are examined and compared taking into account the required characteristics.

Another case study, instead, concerned a refresher course for designers, carried out in the framework of AIPI (Italian Association of Industrial Designers). Participants, all professional designers, were invited to put forward functional schematic diagrams as shown in Fig. 8.13. The theme involved a machine that could move a mass M along a horizontal length

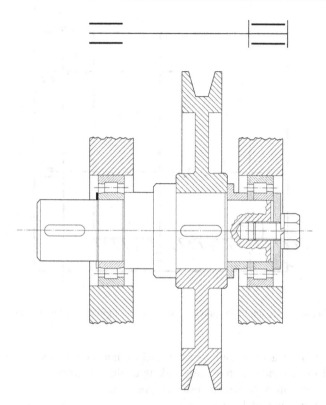

Fig. 8.12 Shaft supported by two cylindrical roller bearings

L and lift it by a height H. The participants, either individually or in groups, developed a diagram, some examples of which are presented in Fig. 8.14, which were then examined and critically discussed together.

A third study presented here is based on experience acquired while teaching the course "Progettazione assistita (Aided design)" at the Università di Pavia [13]. Students, divided into small groups, studied a real machine: each group was assigned to a specific sub-group of the machine, according to the following steps:

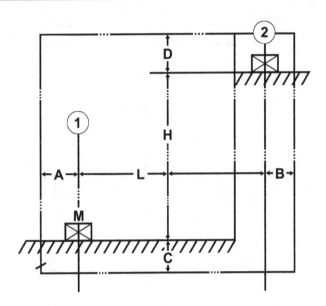

Fig. 8.13 Design theme proposed at a course for professional designers

(a) physical disassembly of the machine into sub-groups
(b) free-hand sketch of parts making up the sub-group
(c) inventory of the dimensions of the parts
(d) virtual model of each part
(e) virtual assembly, creating a virtual model of the sub-group
(f) in groups, a virtual assembly of the sub-groups, creating a virtual model of the complete machine
(g) study of potential modifications of the machine, up to the virtual models of the machine thus modified.

Each of the three case studies presented revealed good results and, most importantly, developed a passion in students to study innovation through creativity.

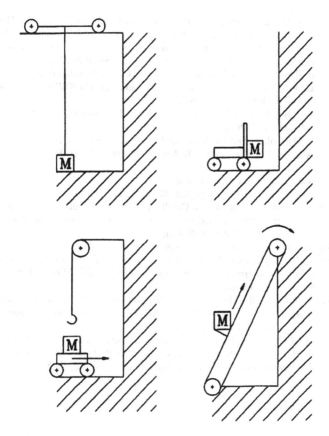

Fig. 8.14 Some examples of principles that emerged from the projects proposed by participants in the above-mentioned course

References

1. https://it.wikipedia.org/wiki/Euristic. Last visit 1 May 2020
2. G.F. Biggioggero,   S. Calabrò, G. Menzio, E. Rovida, *Evoluzione storica di soluzioni costruttive. Il caso di sospensioni per autovettura* (Politecnico di Milano Dipartimento di Meccanica e CNR Progetto Finalizzato Beni Culturali Milano, 2003)

3. S.G. Bona, G. Genta, G. Mimmi, C.E. Rottenbacher, E. Rovida, The ancestors of the motor car: first steps in terrestrial locomotion, in *VI Convegno di Storia dell'Ingegneria—Napoli 2016 2nd International Conference on the History of Engineering*, Naples, Italy (2016)
4. E. Rovida, R. Viganò, A. De Crescenzo, D. Raco, Development of innovative principles to perform given functions, in *AEDS Workshop,* 3–4 Nov 2005, Pilsen Czech republic
5. G.F. Biggioggero, E. Rovida, *Metodi per la progettazione industriale* (McGraw-Hill, 2003)
6. G. Altshuller, And suddenly the Inventor appeared, in *The Theory of Inventive Problem Solving* (Technical Innovation Center, Massachusetts, 1996)
7. www.etria.net. Last visit 10 Jan 2021
8. https://www.aitriz.org. Last visit 10 Jan 2021
9. E. Rovida, M. Bertoni, M. Carulli, U. Giraudo, Integrating TRIZ and QFD effectively in product development: a case study, in *TMCE Symposium,* 21–25 Apr 2008, Izmir, Turkey
10. M. Asimov, *Introduction to design* (Prentice Hall, 1962)
11. F. Rosa, E. Rovida, Development of creativity in the engineering education, *35th International IGIP Symposium with IEEE/ASEE/SEFI 2006–09–18*, Tallinn Estonia
12. F. Rosa, E. Rovida, *La storia dell'ingegneria e la formazione degli Ingegneri,* 1° Convegno Nazionale AISI (Associazione Italiana di Storia dell'Ingegneria) Napoli, 8–9 Mar 2006
13. G. Galli, M. Guarneri, E. Rovida, Experiences of engineering design education at Pavia University, in *International Conference on Engineering Design (ICED 05),* Melbourne, 15–18 Aug 2005

Human Aspects of Technology

9

Technology is for the mankind and not mankind for technology.

9.1 Preliminary Concepts

Scientific research is an activity that, even in its embryonic form, man has performed since the beginning of time. Man has always attempted to explain the natural phenomena he witnessed, and soon started formalizing them.

On the other hand, man immediately provided evidence of his technical skills by creating artefacts to help him perform his daily activities.

The two activities, "scientist" and "technician", were initially totally disconnected: scientists carried out their research, while technicians created their artefacts with little or no consideration for the results of research. Carpenters in Ancient Greece, for example, could build excellent ships but, most probably, did not know Archimedes' principle, and could not rationally explain why ships float. However, they knew how to build them well, based on experience.

Hence, science and technology are almost entirely distinct, and will remain so until the Renaissance, though some hints to their unity do appear earlier; for instance, *Ars sine scientia nihil est* (Art without science is nothing) is the motto of the builders of the Duomo in Milan, Italy.

© Springer Nature Switzerland AG 2022
E. Rovida and G. Zafferri, *The Importance of Soft Skills in Engineering and Engineering Education*,
https://doi.org/10.1007/978-3-030-77249-9_9

Only during the Renaissance will technology start using scientific results, and science will use instruments created by technology. Thus technology was born, and progress started accelerating through the developments of the Seventeenth and Eighteenth centuries, in a dizzying crescendo until the present day.

For example, progress in clockmaking triggered the need to deal with and delve into dynamic problems, for example relative to oscillation, thus further improving clocks.

Another example is the improvement in suction pumps, especially for the extraction of water from mines, which led to the acceleration of studies on problems concerning vacuum, generating further improvements in the construction of pumps.

9.2   Ethical Issues

Scientific research is, in itself, neither good nor bad. Starting from Galileo, it is the process of "reading" the great book of Nature and, therefore, has no specific ethical features. Ethical aspects concern the way it is carried out. For example, research that, despite broadening knowledge, required the sacrifice of human lives would not, of course, be acceptable.

Technological development increases the power of man and is, therefore, ambiguous. Here, accordingly, lies the need for an ethical evaluation tool.

Technological progress can be examined from many points of view that can, nonetheless, be traced as an initial approximation to the following three ideologies [1]:

1. The ideology of indefinite progress, which starts from the consideration of the cumulative character of progress, thanks to which humanity would proceed towards ever better life statuses, in a limitless progression.
2. The political-revolutionary ideology, according to which progress would lead to one class' domination, which could only be opposed by a new regime of the oppressed class.

3. The critical-catastrophic ideology, according to which progress would irreparably compromise the world and would lead to total alienation of mankind.

All three points of view identify, however extremely, an actual factor, but all three also have their weaknesses. Thus, the ideology of indefinite progress is refuted by many current problems; the political-revolutionary ideology is open to the criticism that violent methods substitute oppressive situations with other oppressive situations, while the critical-catastrophic ideology ends up sanctioning an exasperated fatalism.

However, each of them is characterised by the absence of ethical aspects. Hence the need for a conscience, which aims to identify the dangers of technological development, underlining, however, the positive points. One necessary consideration is that technological development is a fact: it is impossible to go back as "de-invention" cannot exist, as history can confirm.

Technological development, as with everything, has both negative and positive aspects but he has brought great benefits to humanity, both in quantitative and qualitative aspects to life. However, specially in the last century, the technology, used in a not always correct and balanced way, has ended to creating problems of environmental impact. The problem is not solved by "going back" and cancelling the progress, but by introducing the concept of "sustainable progress" of technology. In other words, the technology, correctly used, also provides the means to prevent and combat the environmental impact.

Beyond these considerations, positive and negative aspects of technology are now mentioned.

9.3   Positive Aspects of Technological Progress on Mankind

Technology, precisely because it is based on the application of scientific principles, is an encounter with the laws of nature. It is worth recalling Dante's *terzinas* [2] from the XI canto of the Inferno in the Divine Comedy

> He said to me: "Philosophy, to him who hears it
> points out, not in one place alone,
> how Nature takes her course
> from the Divine intellect, and from its art.
> And if you note well the physics,
> you will find, not many pages from the first
> that your art, as far as it can,
> follows her, as the scholar does his master.
> So that your art is, as it were, the grandchild of the Deity.

The concept expressed by the Great Poet is very clear: Nature, as a creation, is a child of God and Art (intended, according to Greek etymology, as the production of objects by man and, therefore, also as "technology") is a child of Nature, as it is inspired by it, and must, therefore, be considered as the "grandchild" of God.

Technological development allows to create machines that can relieve humans of some of their work. By carrying out some of the labour that should have been done by humans, they "grant" humans free time, which can then be devoted to cultural elevation. If then man, as unfortunately happens only too often, uses it to debase himself, it is not technology's fault!

The great progress of modern medicine would not have been possible without the development of technology: suffice to think of machines and auxiliary instruments for medical professionals. It is not by chance that many Engineering faculties have held degree courses in Biomedical Engineering over the span of very few years. Technology, moreover, can now provide aid to disabled people, who only a few decades ago would have been condemned to be excluded from many activities. One of the authors recalls that years ago, climbing the stairs of the Department, he had to make an effort to keep up with a Ph.D. student: that was no surprise, as the student was one third the age of the author, who was nevertheless surprised to find out that the student had an artificial leg!

9.4 Negative Aspects of Technological Progress on Mankind

If it is not well managed and, especially, not filtered by an appropriate ethical spirit, technological development can also generate negative issues. We find such an example in the great availability of material goods that, thanks to technological development targeted at excessive consumerism, can drive some individuals towards materialism, towards the excessive accumulation of material goods. It encourages them to consider themselves and others based only on what they possess, in other words, a rationale of having rather than of being prevails, even to the extreme case of idolatry of material possessions.

Moreover, the availability of some material goods to people who are not able to use them correctly can cause negative situations, even very negative ones: for example, cars capable of very high performance can be the cause of great trouble in the hands of people who are reckless and/or incapable of understanding their limitations.

An example from one of the author's professional experience concerns a vehicle parked by the driver on a slight slope. At a certain point, the vehicle started moving on its own and the driver stepped in front of the car in an effort to stop it. Needless to say, he was run over with dire consequences. If only he had considered the principle of energy conservation and, instead of trying to stop it like that, had jumped back into the car: the injury could have been avoided.

This episode confirms the need for users of technological products, in other words, everyone, to have a scientific-technological background, even at a basic level.

Those mentioned above are examples of a questionable or even negative use of technological development, and involve, especially, but not exclusively, the users of technological products.

There are also inherent errors in the creation of technological products, and these especially involve technicians. For example, hastily designed products, perhaps due to superficiality or the urgency to launch them on the market, can create serious problems.

One of the authors came across this case in his professional practice. A hoisting machine, due to the hurry to release it on the market, was built by literally "copying" a competitor's machine. However, due to haste, it had not been perfectly copied, and the machine thus produced experienced the breakage of a part that, in turn, caused a serious injury.

Another example from the professional experience of one of the authors involves a high performance vehicle, whose engine compartment hood suddenly opened while in motion, impeding the driver's visibility. Only the driver's composure and skill, as well as the slow speed at which he was travelling, prevented an accident. A technical inspection in which the author took part showed that two components, which should have been connected by a screw within the locking mechanism of the engine compartment hood, were not connected at all. The report stated that the screw connection lacked the anti-slackening device and, therefore, it had become loose due to the motion-related vibrations. It was not possible to ascertain if the lack of this device was to be attributed to a design error or to an assembly error.

9.5   Technological Development and Environment

Technological development must still contend with several detractors who are, often, the most avid users of technological products.

Some considerations of technological development can be made on the cycles of the product represented in Fig. 9.1. On the top, is represented the linear economy: without recycle, the accumulation of waste becomes bigger and bigger. The figure below, instead, is derived from the Asimow cycle [3], which was already mentioned in other parts of this book (Fig. 8.9). It

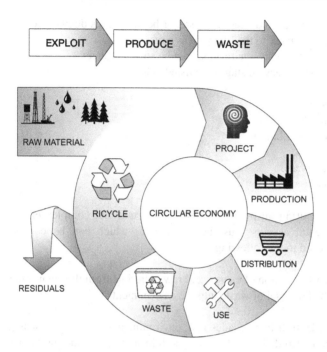

Fig. 9.1 The lifecycle phases of the product in the linear and, respectively, circular economy

concisely represents the life of a technological product, by highlighting the concept of circular economy.

This starts from the project phase, which, by utilizing raw material, consists in the formalization of all necessary information to manufacture the product.

The project "enters" the production phase, which can be considered as constituted by the realization of the parts and the successive phase of assembly.

The product, after the distribution, now in contact with the user, and having become a product in use, begins to fulfil the function for which it was designed.

The use phase, however long or short, still has a time limit and the decommissioned product can follow two paths: be eliminated and become waste, or be, to various degrees, recycled and go back to the lifecycle as a resource.

In this way, the residual can be strong reduced.

The lifecycle stages are listed below:

(a) Project
(b) Production
(c) Distribution
(d) Use
(e) Waste
(f) Recycle.

During all of these phases, important and complex environmental aspects must be considered, which can be concisely traced to the following:

(a) Collection of energy and raw materials from the environment
(b) Disposal of waste into the environment.

Until a few decades ago, environmental issues were scarcely considered. Today, however, both insiders and wider public have started focusing on these. Notwithstanding that going backwards in time is impossible and that a return to a pastoral society is pure fantasy, technology also offers tools and methods for sustainable development. One of such tools is "Design for X", an area of "Design Science" that studies all criteria to face the "X" need. Based on the meaning attributed to "X", we may have "Design for Production", "Design for Assembly", "Design for Utilization" and so on. A particular type of "Design for X", "Design for Sustainability", aims to study the criteria and the measures, both Soft and Hard, which can minimize the above-mentioned two problems (a) and (b). In particular, in terms of item b), a particularly important aspect is the disposal of end-of-life products, which can, principally for some categories of products, involve especially delicate issues.

Further considerations on the bond between technological advancement with environmental topics can be deduced from the

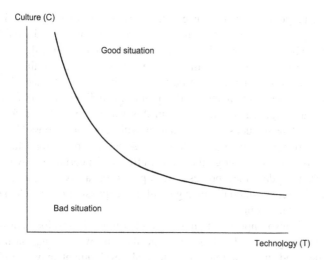

Fig. 9.2   The sustainability hyperbole

diagram in Fig. 9.2. The depicted hyperbole divides two fields, with the field of sustainability located above the curve, and the field of non-sustainability below the curve [4].

The meaning of the curve is clear: to have environmentally sustainable situations, it is necessary that at least one of the two variables should have a high value, but it would be best if that were the case for both variables. The two variables that constitute the "recipe" for sustainability are technological development and culture, intended as incorporation of knowledge and sense of morality.

It is clear even from some prior considerations that technological development is also essential to sustainability.

For what concerns the moral sense, we refer to Chap. 5.

It is worth making some considerations on the need for knowledge. It is very important for an assessment of all actions and of their consequences, as mentioned in the above Chap. 5. One example is deemed especially effective. Anyone who ignores the concept of "food chain" would probably assess the discharge of toxic substance A in the water exclusively from the perspective of damage to the living water organisms and would, probably though arguably, not consider it a moral fault [5].

Let us suppose, now, that this person is informed on the concept of food chain and that he finds out that toxic substance A is noxious also to humans. At this point, this individual knows that aquatic organisms, which live in this water, are part of a food chain that includes humans, and it will thus not be possible to consider the discharge of substance A as free from moral fault. Thus individuals judge the ethical effects of a certain behaviour based on data they possess: the opinion on a given topic can, therefore, substantially change with the expansion of knowledge on the same topic.

It is also important to note that a harmonious development of technology and of culture can contribute to overcoming, on the one hand, the fear that technological development may "steal" jobs, and on the other an uncritical faith in technological development ("Why think? All you need to do is press a button and it does everything for you!"; we have heard this concerning computers).

Concluding, we can say that what is required from technicians is:

- Expertise
- Great attention to all life phases of technical products
- Awareness of the ethical dignity of technological work and a consistent behaviour
- Humility, coming from the awareness that one must constantly study to be up to date
- Opposition, as much as possible, to the use of technology solely for material benefit.

And what is required from users of technological products is:

- Scientific-technical background knowledge
- Knowledge of technical products
- A sense of responsibility
- Awareness of one's limitations (self-criticism).

References

1. C.M. Martini, *Conferenza al Collegio degli Ingegneri di Milano,* 28 April 1984
2. D. Alighieri, *La Divina Commedia. Inferno* XI (97–105)
3. M. Asimov, *Introduction to design* (Prentice Hall, 1962)
4. E. Manzini, *Design per la sostenibilità ambientale* (Zanichelli, 2007)
5. M. Torchio, La bioetica un ponte per la sopravvivenza. Soc. It. Sci.Nat. **65** (1973)

Talent Development **10**

Balance between the intellectual quotient and the emotional quotient.

10.1 General Principles

Talent can be defined as a mainly innate quality, attributed to a natural inclination towards dealing with problems and situations. Talent is often aimed towards a specific discipline. In this sense, we can refer to talent for mathematics, talent for music, talent for business. We are born with the necessary traits to turn lack of preparation into effective skills, so it is possible to refer to talent as innate, and often this is the case. This does not mean that talent cannot be taught, at least up to a certain point. This is precisely the scope of the chapter and, in a sense, of the whole book. Talent can indeed be considered, from some perspectives, as a sort of "integration" of various Soft Skills and, as such, it can be developed and taught to a certain extent. Any constructive action, any life experience, any practiced activity (so long as these activities are critically analysed and interiorised) are steps that bring us closer to our talent development.

The equivocation of talent as inborn: why do we believe this? Because people who often stand out and are successful start devoting themselves to their passions as children, often pushed by parents and/or teachers who support or indulge them.

© Springer Nature Switzerland AG 2022
E. Rovida and G. Zafferri, *The Importance of Soft Skills in Engineering and Engineering Education*,
https://doi.org/10.1007/978-3-030-77249-9_10

This spark is often wrongly mistaken for an innate inclination. But it is actually possible to learn new abilities after growing up. Consider how Leonardo da Vinci was an apprentice to Andrea del Verrocchio: how can we be sure that this time spent carefully and critically observing the gestures and the words of the Maestro did not influence the development of Leonardo's already noteworthy innate talent?

Moreover, innate talent equivocation comes from the idea that a talent is not expressed the same way in each person. The difference between someone who remains a good executor and someone who becomes a "genius" is not genetic or, at least, it is not always so. What decisively affects and contributes to forming talented people is studying, hard work, practice, abnegation, application and discipline. Talent is not unleashed on its own; it actually needs to be cultivated constantly, day by day.

The development of talent can take advantage of certain strategies, which are very useful to fully exploit our innate and unexpressed abilities.

Some of these, which are deemed particularly sound, are listed here:

1. Observe and carefully analyse who we want to become: "nobody is better than anybody else, if someone has achieved their goal, we can do so too". It is important to be inspired by people we admire, carefully considering their work. Note that one should not be blindly inspired by someone who is respected or admired, "copying" their ideas and attitudes without any critical perspective.

2. Emulate top performers, imitating their techniques, their style and the information required to increase the abilities at hand. Pablo Picasso once said: *"Good artists borrow, great artists 'steal'"*. It is very important to carefully, critically and ethically select the top performer chosen to be emulated: for example, a morally questionable model should obviously not be chosen! Moreover, in this case too, emulation should always be critical and, especially, flexible: we need to be open to change model and/or direction, if necessary.

3. We need a *mission statement*, and it is vital to constantly consider this: it is very important to look inside ourselves, analyse ourselves, listening to the opinions of reliable and competent people and deciding what we want to do. It is possible to change one's *mission* but this should not happen too often, wasting time, money and motivation and broadening the distance from the objective. In this regard, see the chapter on *Flexibility*: flexibility does not mean being a "weather vane".

4. Being open to change and to risk, considering mistakes as a source of learning and not as a failure: of course, this means that mistakes should be analysed critically and, in particular, that we understand the *why* and the *what* of the mistake. It is, therefore, very important to pay attention to the mistakes we make, not trying to "hide" them, but analysing them and facing reality in order to avoid further mistakes in the future.

5. Choose austerity to have a wider motivational boost. Many successful people came from challenging backgrounds: poverty, problematic families and various hardships marked their childhood. It is precisely the initial austerity of their life that, despite not being chosen, acted as a trigger for them to emerge, to become "someone" and, often, a "number one". R. L. Montalcini, S. Jobs, W. Churchill, B. Gates, R. Federer, are some examples.

6. Assess the type of technical skills required and acquire them. *Conditio sine qua non* (Indispensable condition) to achieve a good result is to have the necessary skills to achieve said result. Many of the people mentioned in item 5, despite their difficult beginnings, worked hard though the problems, studied, perhaps at night, and achieved a qualification that allowed them to approach life with an edge. Skills, after being acquired and worked on critically, must also be applied. Hence the importance of practicing alone to develop and improve discipline. Talent is inextricably tied to practice. Without constant practice we run the risk of being frustrated, and talent almost always remains an unfulfilled promise. Talent is a seed that needs daily watering.

7. Be capable of dividing objectives into stages. Each goal, especially if it is particularly challenging, should be "sectioned" into sub-objectives that are individually easier to achieve. This will also provide the non-negligible advantage of avoiding frustration without impairing self-confidence and self-esteem.

8. It might be necessary to fight. The "so-called easy" goals have a relative value. To achieve an important goal, it is important to do our utmost.

9. A very frequent error is to seek the easiest solutions. It is an attempt to avoid leaving the personal "comfort zone". One of the authors, at a young age, asked a teacher what should be done to achieve success. He expected an answer like, "Do what I am doing". instead, the teacher answered: "Try to do the most difficult things, those which are generally avoided". The advice is logical, if we consider it carefully. Indeed, by choosing this path, there will be less "competition" (few people want to take on "difficult" jobs) and, additionally, if we do not become very successful, we shall less likely be criticised for it. Finally, the most challenging activities are an opportunity for growth, both in terms of character and of learning a method.

10. It is essential to think with images. Transform the figured concepts in our mind. It would be even better if the images were dynamic, creating a mental movie to stimulate our unconscious mind. This might help to both visualise objectives and to define the most appropriate path to achieve them.

11. Exchanging bad, old habits with new and positive ones. It is a strategy of great importance. We all tend to do what we have always done: laziness, fear of failure when doing something new, and a tendency to avoid leaving our "comfort zone" are all justifications that are often used as an excuse for not changing. This means "becoming hostage" to old habits. This is one of the main obstacles to talent development. We need to keep in mind that leaving the "comfort zone" often means stepping into the "growth zone".

12. To learn something well, we must be capable of teaching it well. This is a great truth, and the gurus of communication say that we remember:

    (a) 10% of what we read
    (b) 20% of what we hear
    (c) 30% of what we see
    (d) 70% of what we see and hear
    (e) 80% of what we personally experience
    (f) 95% of what we teach other people.

    It is interesting to observe how the quantity of what we remember drastically increases when there is an active involvement of the subject (personal experience and teaching others).

    Teaching others (and the authors can confirm this through their personal experience) is a great way to formalise one's own knowledge and thus ideas that one thought were clear but which need some fine tuning, concepts that need developing, mental connections that need to be reinforced can emerge.

    It must also be said that teaching others is a great form of input for our communication skills, and communication is one of the main Soft Skills.

13. If you are stuck, make a change. When facing a challenge (and who has none to face?) we can, in some cases, be stuck and throw in the towel, saying (or thinking) "I will never be able to do this!" Nothing could be further from the truth. When faced with such a situation, we need to analyse the problem, asking some specific questions: "What is exactly the difficulty?", "Where does it come from?", "What are the surrounding circumstances?", "What can I do to overcome this difficulty?", "What can I do to avoid the difficulty?"

    It must, however, be said that such an attitude initially requires a critical analysis of the difficulty. Some issues, objectively speaking, cannot be overcome. Indeed, by no mere chance did the ancient people wisely say *Nemo ad impossibilia tenetur* or *Nobody is obliged to do impossible things*. It is, therefore, necessary to first analyse whether the challenge is truly insurmountable and, if so, then changes must be made, without ending up like Don Quixote who challenged windmills to duels!....

14. Being brave brings together passion, perseverance and self-discipline, allowing us to express additional performance. Courage is crucial, as it is essential to overcome challenges. And once again, as always, we need balance, which was discussed in Chap. 7 "Awareness". If courage is a virtue, illogical and irrational courage, which drives us towards useless and serious risks, becomes "madness".

15. Last but not least, or rather, first of all, we find Intellectual Fusion. It is crucial in order to reach a variation in behaviour. Besides the development of the IQ (Intellectual Quotient), a high EQ (Emotional Quotient) is also important. We shall thus have an individual who can be a "Difference Maker" (DM), a complete individual who can face hard problems with IQ but who can also see problems from a different perspective, through EQ. The combination of the two factors completes Intellectual Fusion (IF):

$$IQ + EQ = IF = DM$$

This will make a professional capable of facing problems in a different way, capable, in other words, of "making a difference".

The development of talent has one great enemy: inability. This can have many different facets, some of which are listed below.

One facet relates to the solution, which can refer to a problem or situation of any type or level. It can have various aspects:

(a) The inability to find the solution, or in other words, not being able to understand what must be done. It has happened to all of us, and it happens often, that we do not know what to do when faced with a problem or a specific situation. We do not know what is better or worse, what the pros and the cons of the various alternatives are. And here a Soft Skill can come to our aid, specifically Imagination and Creativity. It is pre- cisely imagination and creativity that make us think of the various solutions to a problem and the relative pros and cons to each of these, almost as if they were spawned from our mind thanks to imagination.

(b) Inability to apply the solution or, in other words, know- ing what to do, finding the solution but not knowing how to apply it because "it has never been done this way". We inevitably become hostage to habit, to what has been done, incapable of leaving the "comfort zone" and of facing new uncertainties. And here again a Soft Skill comes to our aid, namely Flexibility. By properly applying this Soft Skill, we overcome on the one hand the fear of novelty, of what we have never tried to do, and on the other hand the frenzy to change at all cost, without properly assessing what we might go against. In other words, a certain amount of healthy fear that keeps us away from danger, but not an untamed psychosis.

(c) Inability to sustain the solutions: once the solution has been found and immediately applied, we feel incapable of per- severing on the chosen path. Laziness, tiredness and incon- stancy are enemies that stop us from going on. How often do we see young people enthusiastically choose a degree course and then, after a while, drop out because they believe that "the other one is easier" or "with the other degree I can earn more". Thus they change degree course or drop out com- pletely, wasting time, motivation and financial resources, and increasing their level of frustration. Here again Flexibility, a Soft Skill, together with Awareness, can be helpful: ability to change and flexibility are great, but once a decision has been made, one must follow it to the end, without being thwarted by the inevitable challenges.

(d) The inability to sustain the side effects of the solution. Every solution, to any problem, does not only have highlights but also grey areas. Every perspective, no matter how viable, car- ries side effects. Here we must give a "weight" to the posi- tive and negative effects of each identified solution and, thus, define a balance statement that can direct towards a motivated and rational choice.

A second facet involves the reaction we can have towards a new situation or a sudden problem. The reaction can have various aspects:

(a) The inability to react, when the constant fear of making a mistake stops any behaviour and any action, and the subject remains inert. It has already been mentioned when discussing more than one Soft Skill, how fear, healthy and rational fear, is a positive attitude. This, in fact, helps us perceive and assess risk and, if necessary, helps us avoid it. This type of fear, therefore, belongs to the conservation instinct. Another matter is psychosis which, often kindled by certain *mass media*, exaggerates risks, sometimes non-existent or if existent, it makes them appear greater and more formidable than what they truly are.

(b) The inability not to react, when the subject loses control and engages in behaviours such as rage, unusual and often violent reactions. This aspect is the reverse of the previous one but, in a certain sense, it is "symmetrical". While the previous was a lack of reaction, this is an excessive reaction. In this regard, see the chapter "Awareness". Loss of control is always very negative, as it can lead to doing things we might regret, which can carry serious consequences. We can also observe that often the fault, at least the moral fault, largely falls on those who behave in a way that provokes these reactions. These individuals would do well to consider some Soft Skills, such as Courtesy, Ethics and Integrity, Positivity and Responsibility. We can observe how, often, they behave in ways that lead to unusual reactions on purpose, to paint the people who are the object of these behaviours in a bad light or even place the blame on them. We could react much better by ignoring such personalities and laughing about their behaviour. A famous journalist used to say: "Those who speak ill of me would speak much worse if they knew what I think of them!"

(c) The inability to perceive things correctly, behaving like an ostrich that hides its head in the sand, is another of these facets. This attitude is often due to the lack of observation and the inability to attribute the right importance to things and events as they occur. Often this means being excessively euphoric when facing an event that is initially considered positive but which then turns out to be negative, and to become depressed when considering an event as negative when it is actually positive. Indeed, it is very important to be balanced in judgement and, especially, to shy away from hasty impressions. How often does an event seem positive only to then turn into a problem, while at other times an event might seem negative but actually reveal itself to be an advantage? Here again is the importance of balance and consideration discussed in the Chapter "Awareness".

10.2 The Development of Talent

Talent is like a sleeping giant that must be awakened. It is always possible to draw water from the well of inventive flair, initiative, and courage. We can fulfil ourselves by completing our own and other people's potential, within our own field. When we create something real, something beautiful and good, no matter how small, we ourselves grow and help the world grow too. If, however, we let ourselves be pray to destructive forces, we tend to deprive things and people of life: *we un-grow and make the world un-grow.*

The following are some significant considerations on talent development.

True talent belongs to those who, beyond awareness, imagination, ethics and action, can "sell trust". They must also be able to motivate, inspire, create and improvise: *know how to turn problems into opportunities.* Talented people, on top of focusing on effectiveness, are also focused on being efficient. They need to be able to make the most of their time, weeding their calendar of non-essential appointments, refusing low priority decisions and requests, knowing how to hold optimised meetings and saving that time for considerations and higher level thoughts. They can best exploit their inclinations and their strengths in order to create and strengthen their key relationships. They also need to be agile when seeking new relationships, and in terms of professional relationships, they need to avoid continually clashing with the same people who are not aligned with today's organisational dynamics that, instead, require aligned adaptation to the rapidly changing needs. Today it is very important to go beyond borders, or to create the best networks with people who belong to a wide variety of functions, geographical locations and Business Units. Thus there is the advantage of achieving a greater focus on networking and of not being intimidated by any setbacks. Moreover, it is important to broaden our horizons to access new information, to stimulate innovation, imagination and relational abilities.

Furthermore, talented people must possess balanced energy, showing competence and value, rational intelligence and emotional intelligence. On top of proving that they can act to create trust, they use empathic listening skills to foster critical thinking in others. Talent must beware of the complacency trap: arrogance about one's own energy can play tricks. We need to be brave enough to view success in less heroic terms and to critically consider the decisions made in the past. This can be a challenging exercise, as one must have the strength and the courage to reinvent oneself in order not to fall into the trap of "*habit*", but it is nevertheless very important.

For example, when one of the authors was in charge of International Marketing for a large company, he repeatedly heard the owner at the time say: *I don't care if you make a mistake, so long as it is something new you tried out, and you are willing to critically accept it.* This is because they were led to think that they would have continued to achieve great results, as had happened thus far, while that statement meant that they would have to continue looking for new operative solutions. Beware of those talents who think: "if it is not broken, don't fix it", and do not sufficiently consider strategy. We must not foster uncertainty on what to change and, consequently, make the process last too long, running the risk of being late.

Talents that survive the complaisance trap are generally capable of positively carrying out a valuable creation phase. They are able to reinvent themselves and their abilities in managing potential crises.

It is, therefore, important for a talent to be proactive instead of reactive, and thus to think of leaving behind a heritage. It is very important for the genuine talent to leave a good impression behind. It is also true that talent must lead a professional to check the possibility of identifying who can be a successor. The author recalls this event: "*When I was hired as CEO of a large American company with the aim of creating a branch in Italy where the company had no offices, the first meeting with the President, in Springfield, Massachusetts, deeply impressed me. After the initial introductions, he immediately asked me to consider my potential successor. I answered a little annoyed that I found it inappropriate, as I had just been hired, to have to immediately think of a successor. He told me that my hiring took my talent into consideration and that he did not want to be put into the position of one day hearing that I would leave the job for an offer I couldn't refuse... Awesome, Jim! He even said that I should choose well, because he was sure that I would do a good job and that a weaker substitute would achieve disappointing results. This did not happen as the company is still a leader in its sector. Basically I was hired and at the same time I had to identify and plan the best time for me to take a step back or even to the side*".

Hope, trust and collaboration are values on which talent must lean in order to face an organisational environment where, unfortunately, to this day, it is still hard to promote and support relational aspects (such as Soft Skills), which are the foundations of collaboration.

Synergy becomes a foundational item for success, as it does not mean bringing together various discomyprding opinions in agreement, but expressing the single perspectives and choosing the "best", considering the single points of view. This is an important ability of talent: finding synergy. While the compromise can be expressed as:

$$1 + 1 = \frac{1}{2}$$

synergy is

$$1 + 1 = 3, \text{ or } 5, \text{ or } 7.$$

To this end, it is necessary for talent to be employed in a guided psychological approach, for example respecting the contributions of colleagues, the willingness to experiment with other people's ideas, to care for the way our actions can impact colleagues' work and the final result.

These behaviours are still, unfortunately, rare, due to the fact that a mentality based on lack of trust in other people and on a veritable obsession with status is still pervasive. Talent challenges the tendency to focus on self to, instead, assess what can be learnt from other people. To overcome psychological barriers, which stop us from working well together, it is necessary to be open to listening and to explore other people's ideas in order to express our own, critically if necessary, and ultimately choosing which to pursue.

Knowing how to listen is the first basic rule of communication (even more so if it is empathic). We often cannot listen because we are anxious about our own performance, as we are convinced that our ideas are better than other people's. Hence, we find ourselves navigating conflicts that might have been avoided, wasting important opportunities for a lack of dialogue, making the people who are not directly involved feel left out: the overall result is the decline of the team's effectiveness.

When, on the other hand, our "ego" diminishes, we give each person the opportunity to understand the situation and to, thus, express their opinion, focusing on the objective.

How can we improve our listening skills? It is important to ask open questions. This increases the counterpart's "curiosity" and helps him express his own ideas. It is, therefore, a matter of developing "active listening", which means stifling the urgency to interrupt or dominate a conversation, and to centre it around ourselves or to find solutions to other people's problems by mainly focusing on the implications of our own words.

In effective collaborations, judgement leaves room for curiosity and people can reach an understanding of how other people's perspectives can be just as important as our own. The difference between active and non-active listening is given by the ability to focus on who is listening and not on ourselves. But listening well is not enough, we must also understand the tone of voice, the body language, the emotions, the point of view and the energy the speaker puts into the conversation. Another important area is how one feels in silence. This is the important role of non-verbal communication.

For many people who love listening to themselves speak, the real problem is that, instead of communicating respect and attention, they end up dominating the conversation, rarely letting the interlocutor speak and often not listening at all, as they are focusing on the fact that "the message is me". It is, therefore, essential to learn to be empathic, in other words, to be receptive towards the point of view of other people, even when we do not agree. It is not easy, but it can provide great results, as honestly understanding the difference between varying perspectives can lead to achieving the "best" solution.

Showing empathy makes other people more open to listening: thus collaboration can move forward in a more fluid way, making it easier to achieve mutual sharing of opinions.

It is also very important to feel comfortable in silence. This does not mean only not talking, but knowing how to communicate care and respect while you are silent, which is quite hard for people who love to listen to themselves speak. A useful exercise is to simply listen without giving negative non-verbal signals. We must remember that it is important to use positive body language when interacting with other people.

It is not easy to be receptive of the opinion of someone we do not agree with, but the result can be extraordinary, if we sincerely adopt the desire to understand the differences between the two points of view. Judgment must give way to curiosity, to allow us to understand that other people's perspectives are just as valuable as our own.

A useful exercise to this end is called "expanding other people's thoughts" or *"leading from inside out"*, which involves asking questions not to tip the scales in our favour, but to help the other person, observing the evolution of the conversation, without forming opinions. This could actually be a decisive step towards synergy, where any creative solution makes everyone feel they can be truly listened to.

Feedback is very important, as it constitutes an important way of enriching ideas. This requires three principles:

1—Listening to all ideas put forward by the interlocutor

2—Being constructive (yes, instead of yes but...)

3—Make the interlocutor look good, make them feel important and place them in the condition to persevere in their activities.

Talent often uses the win-win interaction, which is characterised by the fact that all parts satisfy their interest and no part is harmed.

Talent, based on the application of various Soft Skills that have been mentioned here, is a good tool to create leaders that can make teamwork creative and productive, thus determining a style of life.

10.3   Daily Life

Talent is important in all areas of daily life. It is especially important to integrate talent with the various Soft Skills discussed here. Some examples are listed below:

Quick intuition is needed in any situation, of course without making hasty assessments. The ancient saying *Festina lente* or *Slowly make haste* perfectly expresses the Latin philosophy, which tells us to make haste—not to waste time—but slowly, always thinking and considering each step.

It is essential, as repeatedly mentioned, not to be a "hostage to habit", so we should not say "we have always done things this way", "I've never done this before". This attitude leads to the perpetuation of past mistakes and hinders any "new" decision.

Another important skill is the ability, when facing a problem (even any of the small issues encountered in everyday life), to examine various possible solutions and make a rational choice. It is, therefore, important to leave old habits behind, to identify solutions that have never been tried before.

What skills can help in developing and bringing out talent in everyday life?

(a) Courage: the fear of making a mistake, of other people's opinions, is a very negative break in decision-making, even in everyday life.
(b) The ability to abandon "comfortable situations", in other words, what has been the status quo so far, and, therefore, the ease of facing uncomfortable situations even if they are more rational.
(c) Imagination, or the ability to hypothesise possible developments to an existing situation: new ideas can emerge from this and, consequently, their pros and cons.
(d) Memory or, in other words, remembering past situations that can be traced back to the situation at hand, and the ability to recall decisions made and errors committed: this is the best way to avoid repeating the same mistakes.

10.4   Engineering

Talent is essential for an engineer, and we can say that it starts with the choice of the faculty. In this phase talent is attributed to verifying one's own abilities and their consistency with personal interests. An aspiring engineer must, of course, be interested in technology and have a leaning towards hard science but also, and here Soft Skills come into the picture, the desire to do something useful for the community and, in particular, the ability to interact with other people.

After a reasoned choice, talent throughout the course of studies manifests itself and can be developed by critically examining the various study topics, considering them an opportunity for growth rather than "annoying" impositions, and discovering new cultural interests. From this examination, and by developing his talent as much as possible, an engineering student will be able to choose a field that will be more in line with personal preferences.

An important aspect of talent is, however, to objectively and critically face personal aspirations, on the one hand by using the skills possessed, and on the other by exploiting opportunities offered by the "market". For example, one of the authors, throughout his activity as a professor of engineering, often sees students immediately aim for top level management roles. In these cases, the author immediately tries to tone down the aspirations of his young colleagues and invites them to aim to a more modest position from which, if their talent allows it, they will be able to access higher positions.

Moreover, engineers can apply their talents and develop them throughout the course of their career. First of all, the professional task of an engineer can be identified, in sum, as taking care of designing, producing, distributing, utilising and eliminating technical products. They need both hard and soft skills to carry out these tasks. Hard skills, often with excellent standards, are supplied by university studies. In terms of soft skills, the ones discussed in the previous chapters can be very helpful. Some of them are listed below:

(a) Communication: an engineer must communicate effectively, for example through service orders, guidelines to be understood and passed on, technical reports on surveys carried out on plants or construction sites.

(b) Courtesy: fair treatment of colleagues and subordinates is not a secondary element for an effective work environment.

(c) Flexibility: an engineer must be ready to embrace changes.

(d) Ethics and integrity: the importance of deontology for an engineer requires no further comment.

(e) Interpersonal relationship: an engineer almost never works on his own; therefore, the ability to interact with others is very important.

(f) Positivity: it is important for an engineer to transmit positivity to the people he works with.

(g) Professionalism: it goes without saying that professionalism is important for an engineer.

(h) Responsibility: it is important for an engineer to always be ready to take responsibility and make sure that the people working with him are ready to take their own.

(i) Teamwork: an engineer often works with other professionals, so the ability to work in teams is crucial.

(j) Imagination and creativity: for an engineer, creative skills are an important driving force for innovation.

(k) Human aspects of technology: an engineer is a human being who works for and with other human beings; therefore, he must always keep in mind that technology must be for and not against mankind.

10.5 Teaching

It goes without saying that talent is of the utmost importance for teachers. Here again we notice how a teacher's skills must be both hard and soft.

Hard skills, which are beyond dispute, entail expertise in the field of competence and continuing education in order to transfer knowledge that is useful and never obsolete.

In terms of Soft Skills, the ones discussed in the previous chapters can be applied to the teaching profession.

Courtesy, for example, implies fair treatment of students. Flexibility requires teachers to perceive, even from small signs (non-verbal language is important), the degree of attention paid by students to the transfer of knowledge, and to accordingly make adjustments to the method in real time.

Ethics and integrity are essential, especially if we consider that a teacher should, first and foremost, shape people and also be an example.

Interpersonal relationships, positivity, professionalism and responsibility are so important for a teacher that they require no further comment.

A teacher must, especially when it comes to engineering, often organise and guide group activities, as many of the application activities require collaboration between many students. The ability to manage teamwork is, therefore, a very important Soft Skill.

Imagination, creativity and balance are essential for a teacher. Moreover, especially because they have to form people above professionals, teachers need to keep in mind the human aspects of technology.

But the basic tasks of a teacher include two aspects that straddle hard and soft skills:

(a) Foster passion for what is being taught: passion, authentic passion, is perhaps the main aspect, as it allows to overcome any obstacle.
(b) Provide a problem-solving method.

A teacher who gives students these two elements has already done a great job!

As conclusion, it is interesting to consider the following figures. Figure 10.1 represents the links between individual,

(Talent Different Maker)

Socio Cultural Content

Fig. 10.1 Links between individual, environment and socio-cultural contest to reach the "Talent Difference Maker"

environment and its race in the socio-cultural contest, for the "Talent Difference Maker".

Figure 10.2 represents the talent assessment and what is necessary to reach the "five plus top". To this aim, it is necessary to combine potential vs performances, i.e. how the individual utilize the potential (skills) to achieve performances.

For deeper insight, see indications from [1–15].

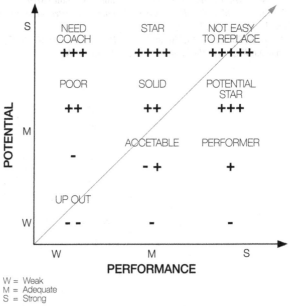

Fig. 10.2 How the individual utilize the potential (skills) to achieve performances

References

1. G. Abbate, U. Ferrero, *Emotional Asset* (Finedit Italia, 2003)
2. A. Brann, *Neuroscience Foil Coaches* (Kogan Page, 2017)
3. T. Cram, *Customers That Count* (Pearson Education, 2001)
4. H. Elrod, *The Miracle Morning* (Magics, 2020)
5. G. Goleman, *Working with Emotional Intelligence* (Bantam Books, 1998)
6. T. Harris, *Io sono OK, tu sei OK* (BUR Saggi, 1974)
7. B. Horowitz, *What You do is who You are* (William Collins, 2019)
8. B. Horowitz, *The Hard Thing About Hard Things* (Harper Business, 2014)
9. S. Johnson, *Chi ha spostato il mio formaggio* (2017)
10. D. Marcum, S. Smith, *Egonomics* (Sperling & Kupfer, 2008)
11. F. Martelli, *Scopri il tuo talento* (Tecniche Nuove, 2015)
12. G. Nicastro, *Alla ricerca del talento* (Way out consultant, 2019)
13. C. Pardini, F. Martelli, *Riconoscere il carattere attraverso l'intuito* (Tecniche Nuove, 2014)
14. D. Pitteri, S. Picucci, R.M. Villani, *Cause Related Marketing* (2002)
15. R. Sharma, *Il Club delle 5 del mattino* (TREGO, 2019)

Proposal

11

Actions introducing soft skills in Engineering Education.

11.1 Experiences of Authors

In this chapter, the authors, starting from their experience, are proposing some actions to systematically introduce the Soft Skills in the University didactic programs. It is important to observe that in the Engineering Education some elements of Soft Skills began already to appear. In this context, the authors have had different experiences. For example, at the authors' initiative, the following seminars were organized, at Politecnico di Milano:

(a) For a Masters student in Mechanical Engineering (Table 11.1).
(b) In preparation for the State exam (in cooperation with the Milan Order of Engineers).
   This seminar, run for many years, lasting 2 h, consists of many arguments of seminar (a).
(c) For engineers involved in the profession (initiative of the Milan Order of Engineers), since 2018.
   The title of the proposed course is "Engineer 4.0. Excellence in the "Intellectual Quotient" is not enough", while the

© Springer Nature Switzerland AG 2022
E. Rovida and G. Zafferri, *The Importance of Soft Skills in Engineering and Engineering Education*,
https://doi.org/10.1007/978-3-030-77249-9_11

Table 11.1 Contents of the seminar (a)

	Subject	Time [min]
1. Presentation	1.1 Engineering 4.0 1.2 Soft Skills 1.3 Communication as a Soft Skill 1.4 Questions and answers	15
2. Communication	2.1 General Principles 2.2 Structure of communication 2.3 Written Communication 2.4 Oral communication 2.5 Questions and answers	90
3. Other Soft Skills	3.1 Engineering Deontology 3.2 Human Aspects of Engineering 3.3 Emotional "Checking Account" 3.4 Development of Personal Acumen 3.5 Questions and answers	90
Total		195

subtitle is "The upgrading of Personal Acumen, through the upgrading of Soft Skills".

This course comprises three stages: the first one is called "Upgrading Personal Acumen", the second is "Upgrading of Social Soft Skills", while the third is called "Upgrading of Organization and Methodological Soft Skills".

11.2 Proposal for Engineering Education

It is now necessary to upgrade the relative education in all steps of Engineering Education: Table 11.2 shows the proposal of general contents related to Soft Skills in the different phases of Engineering Education [1–3].

Table 11.2 Initial proposal of general contents related to Soft Skills in the different phases of engineering education

Degree	Specific courses	Aspects to take into account in the design of technical courses
Orientation	2 seminars: Methods of study and Characteristics of the different Engineering courses	(1) The ability to study efficiently, i.e. an "instruction book" of the brain: how to study for valid results (2) Motivated choice of an engineering field and, consequently, of an engineering course
Bachelors	1 seminar/year to choose from: (a) Communication (b) Psychology of work (c) Ethics	(1) Ability to approach to the structuring a written and oral presentation (2) Some fundamental concepts about the engineering deontology and about the team work, included the psychology of work environment
Masters	1 course or seminar/year to choose from: (a) Creativity (b) Psychology of work (c) Talent development (d) Ethics	(1) Ability to imagine and create new concepts and new ideas in the specific field of activity (2) Ability to realize a positive work environment, by efficient application of interpersonal relationship (3) Ability to develop one's talent (4) Ability to deepening some deontological aspects
Ph.D.	2 courses about upgrading Communication and Ethics	Thesis on Engineering Ethics, by applying the Communication methods
Continuing education	Specific courses: e.g. about personal acumen	Very strong part of the courses devoted to discussion

11.3 Proposal for Teachers for Secondary School

The authors consider that soft skills training is also very useful for secondary school teachers. To this aim, Table 11.3 shows a proposal of topics and relative contents for specific courses devoted to the Soft Skills for teachers specific, but not only, in scientific-technical fields.

Table 11.3 Soft Skills for teachers for secondary school

Topics	Contents
Communication	The teacher's fundamental task is the transmission of information. Therefore it is very important that a teacher formalizes the contents that he must transmit, analyzes the contents in concepts, gradually simpler, and transmits the concepts with valid rules and examples. It is also important for a teacher to convey a passion for what he teaches: to this aim, it is important to utilize examples to affect the emotional side of students
Flexibility	A teacher must be ready to feel the reaction of students: reduced attention, non-verbal signs, questions. In relation to this, it is important to react in such a way as to correct such signs
Ethics and Integrity	The important role of ethics for a teacher does not require underlining. Commitment to teaching, equity in evaluations, availability in organizational activity are clear duties of a teacher
Positivity	A teacher must transmit positive energy. Encourage students who struggle to follow is very important. But is also important properly judge the pupils who are undeathed
Responsibility	A teacher must always remember that his main task is to train people: he must therefore, first of all, be an example of life
Teamwork	Often a teacher has to lead working groups, for example, for exercises, projects, thesis: therefore, a teacher must be able to identify the talents of each one and entrust, to each one, tasks consistent with these talents

(continued)

Table 11.3 (continued)

Topics	Contents
Imagination and Creativity	A teacher must stimulate pupils' ability to innovate: therefore, a teacher should keep up to date on heuristic methods, which direct them to innovative ideas
Human aspects of technology	A teacher is a man who has to train other men: therefore it must always emphasize the aspects of technology in favour of man
Talent development	A key task of a teacher is the ability to identify and develop the talents of pupils
Empaty	A teacher's ability to perceive pupils' problems and moods is important: the teacher's attitude will have to adapt
Learning to learn	It is important for a teacher to be up to date both on what he teaches and on how he teaches
Leadership	A teacher is, in his field, a leader: therefore it must have the necessary charisma

References

1. E. Rovida, The role of University-industry relation in engineering education, in *IGIP (International Society for Engineering Pedagogy) Symposium* (1992)
2. F. Rosa, E. Rovida, R. Viganò, Some considerations about the non-technical subjects in Engineering curricula, in *37th IGIP(International Society for Engineering Pedagogy) Symposium* (2008)
3. P.S. Percy, Italian engineering deans council, in *Conference at Pavia University* 25 Sept 2007

Printed in the United States
by Baker & Taylor Publisher Services